Urban Geography

Urban Geography

Social and Spatial Dynamics of the City

Barney Warf
Professor
Department of Geography
University of Kansas
Lawrence, KS, United States

ELSEVIER

ACADEMIC PRESS
An imprint of Elsevier

Academic Press is an imprint of Elsevier
125 London Wall, London EC2Y 5AS, United Kingdom
50 Hampshire Street, 5th Floor, Cambridge, MA 02139, United States

ISBN: 978-0-443-30281-7

For Information on all Academic Press publications
visit our website at https://www.elsevier.com/books-and-journals

Publisher: Peter Linsley
Acquisitions Editor: Rakhshan Rizwan
Editorial Project Manager: Ruchi Bhargava
Publishing Services Manager: Deepthi Unni
Production Project Manager: Surya Narayanan Jayachandran
Cover Designer: Mark Rogers

Typeset by MPS Limited, Chennai, India

Contents

CHAPTER 1

Introduction

Chapter Outline

Cities are arguably the greatest creations of the human race. Cities are where civilization began. They generate the vast majority of the wealth on the planet. Cities are where the world's financial capital is concentrated, including banks and investment firms. Their networks define the global economy. They are headquarters to innumerable corporations and they create labor markets that employ billions of people. They are where most of the world's media are concentrated, including publishing, television, and Internet services. Cities create the vast majority of the world's information. They are centers of innovation and creativity, as well as of the arts and sciences. Cities are where political power is concentrated, including governments, politicians, and networks of the elites.

Obviously not everything about cities is wonderful. They can be clusters of intense poverty and crime. Many of the world's problems are urban problems. Cities have enormous environmental impacts, changing land uses, sucking up great quantities of water and energy, devouring agricultural land, and generating most of the gases such as carbon dioxide that are responsible for global climate change.

Cities have been a predominant part of the human geographies of the world for the last 4,000 years or so. They became particularly important in the late medieval period. Cities are where capitalism was born and where the Industrial Revolution began. Modernity began in cities. And cities have been centers where various social and cultural groups interact, mix, and influence one another.

Because cities are productive and dynamic, offering many opportunities, they attract large numbers of people. For the last two centuries, there has been a steady stream of migrants from rural areas to urban ones. Today, for the first time in human history, the majority of the planet's population lives in cities (Fig. 1.1).

This volume introduces readers to the ways in which cities emerged, how they changed over time, and how they function, that is, the social and spatial dynamics of cities. The viewpoint adopted here is resolutely historical: what happens in cities did not materialize out of thin air, but reflects social processes long in the making. The focus is on urban political economy,

Urban Geography. DOI: https://doi.org/10.1016/B978-0-443-30281-7.00001-3

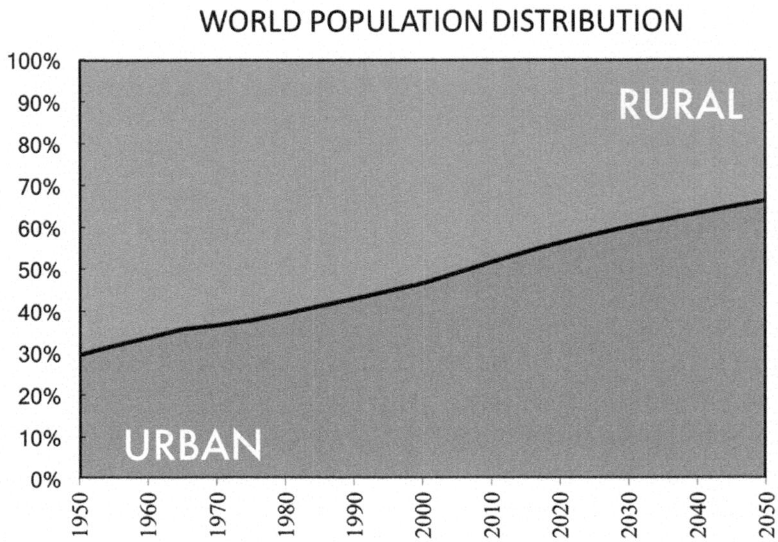

Figure 1.1
Percent of World's population living in rural and urban areas over time.
(https://commons.wikimedia.org/wiki/File:Historical_global_urban_-_rural_population_trends.png.)

which views cities as outcomes of social relations, that is, the economic, political, and cultural forces that are embedded in cities and shape how they are structured and change.

Cities are not random collections of people, in which case they would be impossible to understand. Cities are structured; they have a form and texture that is produced by their role in the broader economy and society. Cities are produced by the decisions, actions, and behaviors of large numbers of people in everyday life. People choose where to work, and in what kinds of jobs and careers, where to live, what to consume, how and where to travel, where to spend their free time, where and what to do for entertainment, and their social networks, such as friends. But people do not have unlimited abilities to make choices: their choices are always constrained by their income, their ideologies, and by the constraints of a society (legal, political, and cultural ones), including racism, sexism, and ethnic discrimination. These constraints mean that different groups of people face different obstacles and sets of abilities, which reflect the relations of class, gender, race and ethnicity, sexuality, age, and other social variables. In addition to people of different social classes, property developers, government officials, urban planners, the legal system, firms, the media, and lending institutions (banks) shape cities in countless ways, including the ownership of land, building codes, fire and safety ordinances, and the availability of credit. Thus, cities are more than just the sum of individual decisions: they have an order and logic that transcends individuals.

There are several dimensions of cities that should be highlighted. First, *cities are historical* in that they are embedded in historical contexts and processes. The landscapes of cities, which continually change, reflect the outcomes of accumulated decisions over long periods of time. *Cities are economic*: under capitalism, they are machines for the production of commodities (goods and services) and the realization of profit. The structure of commodity production shapes the locational requirements of firms and companies, which in turn affects patterns of employment and land use. *Cities are environmental*, meaning that they are shaped by, and in turn shape, the biophysical world around them, including local climates, land uses, and flows of water and energy. Cities generate vast amounts of waste that needs to be disposed of properly. *Cities are political*, meaning that they contain varying social factions vying for power, and are occasionally the sites of intense class and social conflict. *Cities are cultural*, which means that they generate vast amounts of cultural phenomena, ranging from oral stories to books, movies, performances, and street life. All of these dimensions are interrelated with one another, interacting and shaping the urban environment in complex ways.

1.1 Urbanization

Urbanization refers to the growth of cities, the process by which societies become increasingly urban in nature. It includes the rise in the number of urban areas and the expansion of their populations and area. Urbanization has varied in tempo and nature throughout history and takes different forms in different places.

Urbanization has a long history stretching back to the early cities that emerged during the Neolithic Revolution. Following thousands of years of village life, the world's first cities appeared in the Middle East around roughly 4000 BCE, including those in Sumer and Babylonia (e.g., Ur, Uruk, and Babylon), to be followed by others in Egypt, the Indus River valley (e.g., Harappa and Mohenjo-Daro), and northern China. Over time, Chinese cities such as Hangzhou became some of the largest in the world. Several places in Africa developed cities, including multiple West African kingdoms, Axum in Ethiopia, and Zimbabwe. In the western hemisphere, cities arose in Mesoamerica under the Mayans and Aztecs, the Andes under the Inca, and the mound-building cultures of North America such as Cahokia. Temples and religious structures played a central role. Cities in Europe began to flourish during the Bronze Age, including early Greece and, later, the Roman Empire, which spread them across much of western Europe (e.g., Paris, London, Toledo, Cologne, and Vienna). Rome itself may have reached 800,000 people during the height of the empire. In the Middle East, the Arab Empire saw the rise of Damascus, Baghdad, Cairo, and other places. In Central Asia, cities flourished along the Silk Road routes (e.g., Samarkand, Bukhara, Tashkent).

Medieval Europe was not known for high rates of urbanization. However, starting around CE 1000, as agricultural productivity began to rise, cities began to grow and many "new towns" appeared, particularly in northern Europe. Trade and migration flows linked many of these together. Hanseatic League cities formed trading networks that grew in size and density across the Baltic and North Seas. Whereas traditionally southern European cities were the largest on the continent, the gradual emergence of capitalism shifted the locus of growth to the north. Urban areas were often the crucible in which capitalist social relations were forged.

The Industrial Revolution marked a rapid increase in urbanization. The steam engine, including trains and steamships, and, later, the telegraph, played important roles. Urban areas offer agglomeration economies that greatly reduce transport and communications costs among firms. By clustering together, firms are more productive than they would be alone, increasing productivity. As a result, cities generate the vast bulk of the world's wealth. In the process, firms generate employment that attracts migrants. The concentration of jobs in cities and the mechanization of agriculture, which drove large numbers of peasants off their farms, gave rise to several societies in Europe and North America that were predominantly urban. Rural-to-urban migration has been the largest motor of urbanization in many contexts. In Britain, cities such as Manchester, Birmingham, Leeds, and Glasgow became the "workhorses of the world." Britain, the first industrial country, also became the first primarily urban one. Industrialization soon thereafter propelled urban growth in France, the Netherlands, Germany, and Scandinavia. The commodification of land and labor gradually led to a spatial distinction between areas of production and reproduction, necessitating longer commuting trips between the two. Many industrial cities were unpleasant and unhealthy places to live, very crowded, and suffering from poor sanitation and air quality; uncollected refuse, rodents, and contaminated water made many cities breeding grounds for disease. In Britain, Friedrich Engels and Charles Dickens were astute observers of the poor quality of life for the urban working class. Similarly, immigrants flooding to the United States in the late 19th and early 20th centuries often found themselves living in deplorable tenement buildings. In 1920, the United States became a majority-urban country. Urbanization is often portrayed using an urbanization curve, which illustrates a country's changing percentage of urban inhabitants over time (Fig. 1.2). In 1800, only 3% of the world lived in cities; in the 20th century this rose from 15% to 39%; in 2007 it reached 50%.

European colonialism led to a worldwide wave of urbanization. Cities in Europe, such as those in Britain, exploded in size and wealth. Because the maritime world economy played such an enormous role, many colonial cities were located on the coast, often near the mouths of rivers. In North America, Boston, New York, and Philadelphia are examples. In Latin America, the Spanish founded Havana, Los Angeles, San Diego, San Francisco, Lima, Santiago, and Caracas. In India, colonial powers established Kolkata, Mumbai, Goa, and Madras. African colonial cities include Lagos, Cape Town, Johannesburg, Maputo, Dakar, and Freetown. In Southeast Asia, colonial port cities included Singapore, Hong Kong, Jakarta, and Manila, and in Australia, Sydney. Because they were centers of capital investment and jobs, such cities became magnets for waves of rural–urban migration.

Urbanization became a central issue in the social sciences in the early 20th century, leading to numerous studies of cities as economic, social, cultural, and political phenomena. Urban geography and sociology largely emerged in Chicago during the 1920s. Urban theorists

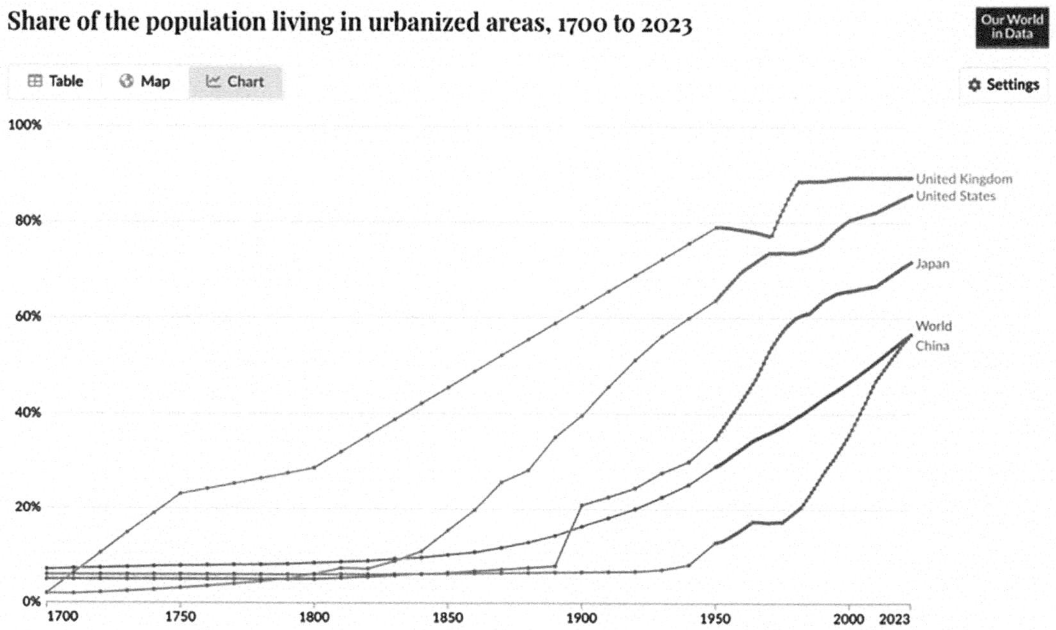

Figure 1.2
Share of population in urban areas, 1700–2023.
(https://commons.wikimedia.org/wiki/File:Urbanization,_1750–2023.png.)

studied phenomena such as the emergence of neighborhoods, their transitions among ethnic groups, and the roles of income and the family life cycle in structuring urban residential space. Others focused on the social and psychological impacts of city life, including crowding, loneliness, and alienation. Cities were often seen as centers of modernity, technological innovation, and cultural change. Conversely, some observers viewed cities as hotbeds of social pathologies, including crime, prostitution, disease, and antisocial behavior.

The state became actively involved in urbanization in the 20th century, building urban infrastructure (e.g., water), housing, and transportation systems. Zoning ordinances were introduced and became ubiquitous. After World War II, the US federal government subsidized housing (e.g., with tax deductions for mortgages) and built the interstate highway system. Urban planning became a major tool of state intervention, rationalizing land uses, controlling negative externalities, and promoting growth.

Globalization has had enormous impacts on urbanization. At the top of the global urban hierarchy sits a handful of global cities, the command and control centers of the world economy, including London, New York, Tokyo, and, increasingly, Los Angeles, Toronto, Paris, Rome, Osaka, Seoul, Shanghai, Beijing, Singapore, and several others. Each of these is the center of headquarters for transnational firms, financial markets, media, tourism, fashion, and other globe-spanning activities.

A definitive characteristic of urbanization in the developed world (and, increasingly, the developing world) is suburbanization. Although its roots go back to the streetcar suburbs of the late 19th century, suburbanization became the dominant face of urbanization in Europe and North America after World War II, when rising incomes and birthrates led many families, especially white, middle class ones, to leave the cities for the peripheries. Many viewed the move as an escape from the perceived high crime rates and taxes, and poor public schools, in central cities. In many wealthy countries the plurality of people live in suburbs. During the post-World War II economic boom, the growth of closely located urban areas fueled speculation about the rise of megalopolises. Low transport costs and affordable housing even led to the rapid rise of exurbs on the metropolitan periphery. Many suburbs moved quickly from being bedroom communities to having commercial centers in their own right. One result was the impoverishment of central cities, which saw their tax bases decline and, consequently, their ability to provide public services. Inner city ghettos began to rise in size and the severity of their poverty, and were largely populated by ethnic minority populations. In the United States, this process witnessed the emergence of impoverished African-American communities; in France, ghettos were populated primarily by people of North African heritage. In the late 20th century, gentrification, propelled by the growth of producer services, saw numerous white-collar professionals take up residence in many urban cores, a process that increased property values and rents at the expense of lower-income families.

Today, roughly 55% of the planet lives in cities, or roughly five billion people, a marked departure from the long tradition of rural life (Fig. 1.3). The definition of "urban" varies considerably. The US Census in 2020 defined an urban area as a densely populated area with a minimum of 2,000 housing units and a population of 5000. The percentage of people living in cities varies widely around the world (Fig. 1.4); generally, urbanization rates correlate closely with levels of national income.

Currently, the vast bulk of urbanization occurs in the developing world, largely due to rural-to-urban migration. Many rural denizens flee areas with low levels of private and public investments, inadequate government services, and few job opportunities, often moving to cities with unrealistic expectations. Most of the world's large urban conurbations are located in the developing world, including metropolises such as Shanghai, Beijing, Mumbai, Delhi, Kolkata, Karachi, Jakarta, Manila, Bangkok, Nairobi, Cairo, Alexandria, Kinshasa, Mexico City, Rio de Janeiro, and Sao Paolo, many of which exceed 20 million inhabitants. Such cities are primate cities within their urban hierarchies, much larger than any other in the same country. Urbanization in the developing world, however, differs from that in the developed, given that cities there generate relatively few jobs, a phenomenon often labeled "urbanization without industrialization." Lack of adequate housing results in enormous rings of shantytowns or slums on the peripheries of many large cities in which more than two billion people reside today, often without sufficient drinking water, electricity, or transportation.

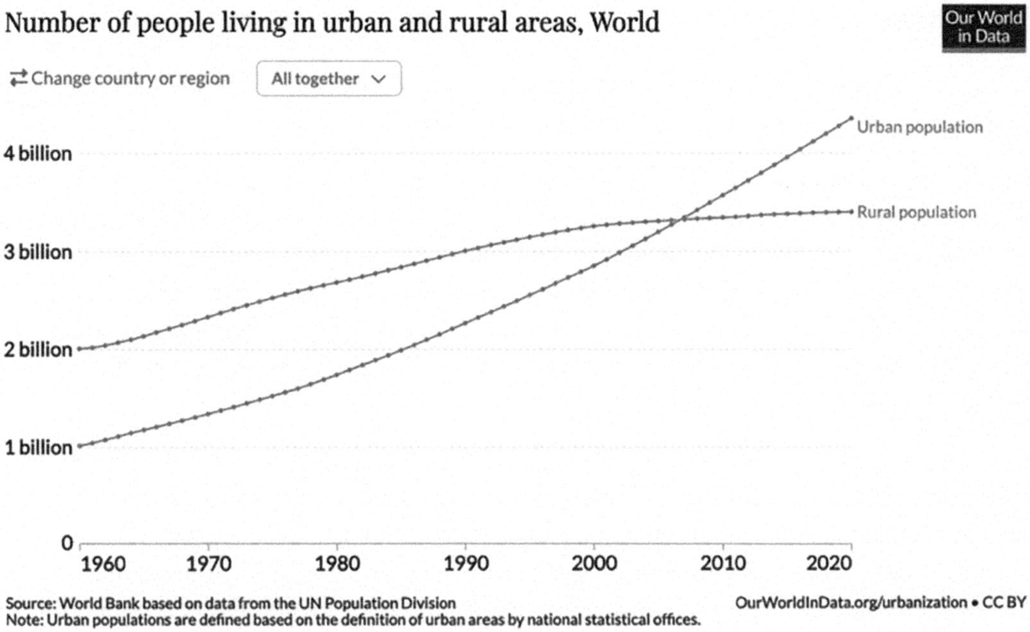

Figure 1.3
Number of people living in rural and urban areas.
(https://commons.wikimedia.org/wiki/File:Urban_and_Rural_Populations.png https://commons.wikimedia.org/wiki/File:2015_World_Urbanization_Map.png.)

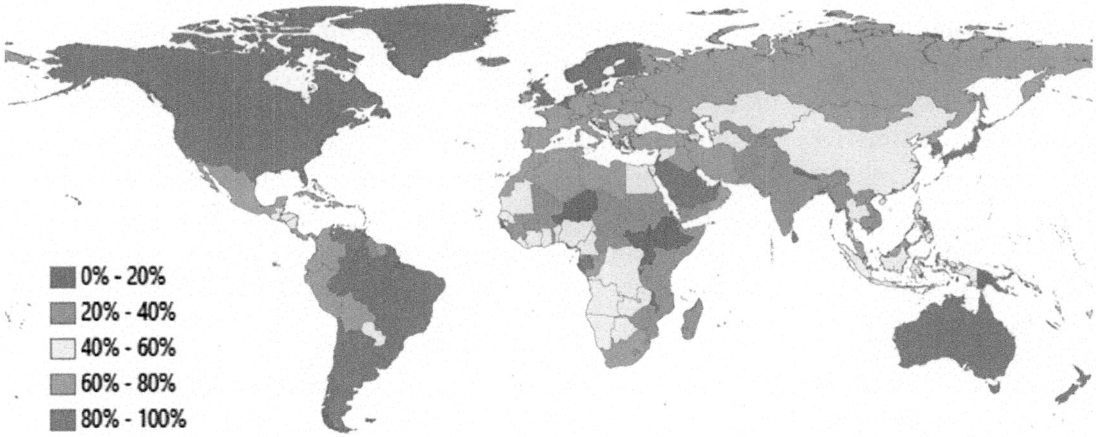

Figure 1.4
Urbanization rates globally, 2015.
Map lines delineate study areas and do not necessarily depict accepted national boundaries.

1.2 Outline of Chapters

The chapters in this book sketch several dimensions of cities as multifaceted, complex, and ever-changing sets of social relations embedded in particular places.

Chapter 2 begins by situating cities in their historical context. In the broad sweep of history, cities are relatively recent phenomena, comprising only about the last 2% of human existence. The chapter notes that only following the invention of agriculture around 10,000 years ago during the Neolithic Revolution did the first villages arise, for urban areas rely on the surpluses created by peasants working in the fields. It traces the rise of cities in several places, including Mesopotamia, Egypt, the Indus River valley, China, Mesoamerica, South America, North America, Africa, and Europe.

In Chapter 3, the historical development of Western cities is considered. Medieval urbanization, while small, was nonetheless important. The chapter considers the explosive impacts of the Industrial Revolution, when mass urbanization began in earnest. It then focuses on various phases of urbanization in the United States, starting with the colonial era and progressing through the 19th and 20th centuries, when American cities were shaped by the rise of industrial capitalism on the continent, waves of immigration, technological changes, and territorial expansion.

The fourth chapter examines the urban economy. It begins by summarizing agglomeration economies, or the benefits that firms derive from clustering together. Nowhere is this phenomenon more important than in downtown areas, or central business districts. It then summarizes urban export base theory, the notion that the export-driven part of the urban economy is the motor of economic growth, bringing in revenues from elsewhere. The chapter also delves into urban land markets, the rings of land use that reflect the most intensive patterns near the central business district (CBD) and gradually decline with distance away from it. More contemporary views of urban economies are explored through a discussion of urban political economy, including growth coalitions. The chapter then summarizes urban economics in light of cognitive-cultural capitalism, the particular form of information-intensive, service-oriented capitalism that dominates cities today, including the role of the "creative class" of white-collar professional workers. Finally, given the widespread impacts of information technologies, it addresses some of the major impacts of telecommunications, including telework and the dispersal of back offices.

In Chapter 5, cities are viewed as systems embedded within systems of cities. It starts with an overview of urban hierarchies, which lock cities of varying sizes together. It comments on small towns, the lowest tier of the urban hierarchy, in which urbanization takes a different form from that found in large metropolitan areas. It briefly summarizes three classical approaches to this issue, including Central Place Theory, the rank-size rule, and primate cities.

Next, the chapter addresses interurban flows of people, goods, capital, and information. It then takes on the issue of city-regions, the broad conurbations of cities and their surrounding hinterlands to which they are closely tied. It concludes with a section on global cities, the command and control centers of the world economy that dominate global economic space.

Chapter 6 takes on the meaty question of urban residential space. It opens on a historical note, that is, the social ecology of the Chicago School, which has had an enormous impact on subsequent generations of thought about how residential areas are constructed. It briefly addresses the question of housing markets, the mechanisms of supply and demand that determine the availability and price of different kinds of housing. It zeroes in on neighborhoods, a fundamental component of urban residential space that reflects the opportunities and constraints faced by different social groups. It then discusses suburbanization, the rapid growth of the urban periphery that is a defining feature of most urban areas today. Next, it concerns itself with gentrification, the growth of relatively affluent areas near the city center that is driven by the labor markets of advanced producer services and the professional jobs that they create. The flip side of urban wealth is urban poverty, as reflected in urban ghettos; given that class and race are inextricably fused, ghettos are predominantly populated by people of color, and suffer all manner of social problems. The chapter concludes by discussing homelessness, the crisis of those so poor and disadvantaged that they are shut out of housing altogether.

Cities are environmental as well as social phenomena. In Chapter 7, various dimensions of the environmental aspect of cities are discussed. It begins by looking at urban climates, specifically urban heat islands. Next it turns to air pollution, a noxious byproduct of urban growth. It then focuses on urban water, including the provisioning systems and the disposal of wastewater. A brief section concerns itself with urban waste management. It also discusses urban green spaces, including parks and green belts. The chapter notes that the management of hazardous waste and access to green spaces are issues of environmental justice. It discusses rural-to-urban land conversion, the process that swallows up agricultural land on the urban periphery. It nods at urban political ecology, which foregrounds environmental issues of the city in a political light. The chapter also addresses hazards of the city, including hurricanes, flooding, and earthquakes. Finally, the chapter points to cities and climate change, the great existential threat of our time.

Chapter 8 makes the case that cities are not simply products of markets but of the state as well. It focuses on the state's role in several ways. Drawing on the specific instance of the United States, it examines national urban policies. It then discusses urban planning, which is hugely influential at the local level. It examines the provision of urban services, especially education and public transportation. The chapter discusses zoning, the reasons it exists, its various types, and impacts. It then examines state intervention in housing markets, including public housing and the thorny debate over rent control.

Most cities and most urbanization today occur in the developing world, the vast set of societies located in Latin America, Africa, the Middle East, and Asia. Chapter 9 begins by emphasizing that cities in the developing world are not some sort of "backward" version of those found in the economically developed world. It next addresses the historical context of urbanization in developing countries, noting the existence of precolonial cities as well as the profound impacts of colonialism. It summarizes the drivers of urbanization in these societies, particularly rural-to-urban migration, a response to the uneven development of capitalism within countries. Next it turns to labor markets in developing world cities, particularly the bifurcation between the formal and informal economic sectors and the types of jobs they create. It then examines residential space in the cities of developing countries, particularly the growth of enormous slums and shantytowns that surround almost every city from Sao Paolo to Jakarta, from Tehran to Cape Town. Finally, the chapter drives home the point that urbanization varies widely among countries in this region, ranging from desperately poor cities to the gleaming towers of Abu Dhabi.

Chapter 10 looks at everyday life and the social geographies of cities. This view is essential to avoid portraying cities as lifeless and inert. It opens with three sections that examine class, ethnicity, and gender, respectively, and the difference that each of these makes. It notes that sexuality is a public as well as private matter, and notes the existence of gay neighborhoods and gay gentrification. Age is another dimension of the social life of cities, as people experience different needs and exhibit different behaviors at various stages of the life cycle. The chapter touches on the role of disability in the city, noting the formidable obstacles to mobility encountered by disabled people everywhere. It briefly acknowledges the role of religion in urban life, particularly the role of churches in shaping community behavior and their role in charitable work. Cities are often portrayed as crime-infested, and some are, although this aspect of urban life is frequently exaggerated. The chapter addresses the city as a playground, noting the vital roles played by recreational centers and tourism. It concludes with comments on the city at night, when most people are asleep but when the night-time economy nonetheless continues to function.

The microelectronics revolution and the rise of the Internet have had profound impacts on urban space. Chapter 11 examines digital cities in five respects. It opens with comments about the fiber-optic infrastructure that underpins information systems in urban areas and welds them together. Next, it comments on some of the impacts of this issue on urban innovation, recreation, and retailing. The third part focuses on urban digital divides, or social and spatial inequalities in Internet access and the implications of this process. Next it addresses urban e-governance, or how cities use information technologies to provide information and services to their residents. Finally, it raises the issue of smart cities, those that deploy information technology in traffic management, energy efficiency, and other realms in the aspiration of sustainable growth.

Further Reading

J. De Vries, European Urbanization. Routledge (2013), 1500–1800.

D. Gollin, R. Jedwab, D. Vollrath, Urbanization with and without industrialization. Journal of Economic Growth. 21 (2016), 35–70.

D. Haase, B. Güneralp, B. Dahiya, X. Bai, T. Elmqvist, Global Urbanization. The Urban Planet: Knowledge Towards Sustainable Cities. 19 (2018), 326–339.

T. Hall, H. Barrett, Urban Geography (5th ed.). Routledge.

R. Jedwab, D. Vollrath, Urbanization without growth in historical perspective. Explorations in Economic History. 58 (2015), 1–21.

A. Jonas, E. McCann, M. Thomas, Urban Geography: A Critical Introduction. Wiley Blackwell (2015).

D. Kaplan, S. Holloway, Urban Geography (4th ed.). Wiley (2024).

R. LeGates, F. Stout, R. Caves (Eds.), The City Reader. 8th ed., Routledge (2025).

X. Peng, X. Chen, Y. Cheng, Urbanization and its Consequences. Eolss Publishers (2011).

A. Smart, J. Smart, Urbanization and the global perspective. Annual Review of Anthropology. 32 (1) (2003), 263–285.

CHAPTER 2

The Origins of Cities

Chapter Outline

Cities, like people, have a birth, life span, and eventual death. Compared to the length of time *Homo sapiens* has lived on Earth (roughly 300,000 years), cities are relatively new in human experience, emerging only within the last 4%–5% of human existence. This chapter traces the rise of cities out of the Neolithic Revolution, which saw the emergence of agriculture. Then it engages extensively with the world's first civilizations, all of which were centered on cities. In several parts of the world, within a fairly close time range, cities appeared on every inhabited continent.

This chapter examines the beginnings of cities in two steps. It starts with a summary of the Neolithic Revolution, when agriculture began, noting how the transition from hunting and gathering to farming created numerous social, economic, and political changes that led to cities. The second part is a long tour of the places where cities began, including Mesopotamia, Egypt, the Indus River valley, China, Mesoamerica, some sites in North America, several places in Africa, and Europe.

2.1 The Neolithic Revolution and the Rise of Agriculture

The origins of cities can be traced back to the rise of agriculture, in which human beings gradually abandoned the hunting and gathering lifestyle that had sustained them for hundreds of thousands of years. Beginning approximately 10,000–12,000 BCE, a widespread series of

Urban Geography. DOI: https://doi.org/10.1016/B978-0-443-30281-7.00002-5

changes began to occur around the world that initiated the end of the Old Stone age, or Paleolithic, and the beginning of the New Stone age, or Neolithic. Central to these changes was the invention or discovery of agriculture, one of the most important transformations in the history of the planet. Agriculture changed how people related to nature and how they related to one another, giving rise to new types of societies, cultures, and landscapes. Essentially, it allowed people to become food producers rather than food collectors.

Agriculture originated in the domestication of plants and then animals. A domesticated plant is deliberately planted, raised, and harvested by humans; a domesticated animal depends on humans for food and, in many cases, shelter. As a consequence, domesticated plants and animals differ from their nondomesticated counterparts. From the human perspective, the domesticates are superior in that, for example, they bear more fruit or provide more milk: they have been deliberately engineered for these characteristics through selective breeding over long periods.

In addition to agriculture, the Neolithic unleashed numerous other profound changes. One of these was the rise of small, settled communities, or villages of varying sizes in which people lived year round. Staying put in one place was necessary to tend crops and farm animals. Thus the Neolithic effectively ended nomadism in vast parts of the world. Many small communities became the centers of increasingly complex divisions of labor, with workshops, houses, religious and ceremonial centers, and the like. Only several thousand years later did some of these begin to emerge into true cities.

Within these villages, artistic, scientific, and intellectual life flourished. Astronomy, for example, became important in many early cultures, not only for religious reasons but to yield accurate dates to predict rainy seasons and floods, the rise and fall of water levels on rivers, and the changes in the seasons. Thus, cultures as disparate as the Mayans in Central America, the Chinese, Indians, Cambodians, and others all developed calendars of varying degrees of accuracy. Similarly, the wheel, likely invented in the Middle East in the sixth or seventh millennium BCE, revolutionized land transportation and increased the velocity with which people, goods, and ideas could travel across space. The oxcart and sailboat did likewise. Agricultural innovations included the plow and the sickle, which raised productivity levels. The invention of baskets during this time allowed goods to be carried and stored more effectively. Evidence of weaving goes back several thousand years as well, allowing people to make clothes that were stronger, warmer, and more comfortable.

The Neolithic also saw the rise of the first class-based societies. Hunter and gatherer groups were typified by high levels of equality, with almost no differentiation among members in terms of wealth. The Paleolithic was, essentially, a period of classlessness, or even communism. In contrast, the rise of agriculture, which generated a surplus social product, allowed a small group of people (typically no more than 5%–10%) to engage in activities other than farming. Agriculture's great advantage over hunting and gathering, after all, was the ability to generate a larger and more reliable stream of food. In the process of creating even a modest surplus, or producing more than they consumed, early farmers laid the basis for the rise of a small elite that came to dominate all Neolithic societies. For the first time in history there appeared kings and queens, emperors, generals, priests, soldiers, scholars, and teachers, all of whom reflected the more complex division of labor coming into being. Conversely, large numbers of people became slaves; indeed, slavery was the world's first form of class-based society, and remained dominant in parts of the world for millennia (e.g., in Europe until the collapse of the Roman Empire in the 5th century AD). Rather than a simple sexual division of labor such as that found in nomadic societies, agriculture gave birth to stratified social systems marked by high degrees of inequality.

Central to the rise and perpetuation of class in the Neolithic was the emergence of the state, or government in all its forms. The state is essentially the means by which a small, elite class manages and controls a much larger population of producers. Early Neolithic societies in many parts of the world developed increasingly elaborate and powerful state systems, including civil servants, religious bureaucracies, and organized militaries. For the first time, warfare as a systematic, organized means of violence among and within societies became a regular feature. With the state, taxes became an obligatory payment, a means of extracting surplus value from producers and transferring it to the elites.

Writing, too, was another innovation of the Neolithic. Many forms of writing began to appear around the world shortly after the birth of agriculturally based societies, including places such as Egypt (hieroglyphics), China (pictographs), Sumeria (cuneiform), the Mayans (glyphs), Phoenicians (who invented the early alphabet Linear A), and the Indus River valley culture of the early Dravidians. Writing was a profoundly important innovation, freeing information from the human mind and the constraints of memory, and allowing it to be stored and transmitted more effectively. For this reason, historians often date the beginning of history to the rise of written records. Only a small handful of people ever mastered this difficult task, including scribes or some priests; most people had no need of it. The most widespread examples of early writing include inventories of goods to be taxed by the state, revealing the close association between class, the state, power, and information. Other applications included commemorations and inscriptions on monuments and tombs, official letters, treaties, and declarations, and the production of literary texts, often mythological in nature and designed to celebrate those holding power. An early example from Mesopotamia was a recipe for beer!

Closely associated with the emergence of early agricultural societies was the development of organized religious systems. Religion, in the form of various animistic beliefs, had long played an important role in hunter and gatherer societies. But with the birth of class societies and the state, religion became much more than a set of beliefs or stories told by local healers or storytellers, and increasingly became an organized, bureaucratized system of political and ideological power. Priests played important ceremonial and political roles, for example, in advising the rulers of almost all ancient civilizations, placating gods and making sacrifices, predicting or interpreting signs of danger (e.g., drought, eclipses, or invasion), and spreading dominant ideas among the populace. Priests were frequently the only literate members of many classical cultures, with access to divine texts. Almost all societies were polytheistic until the invention of monotheism, likely by the ancient Hebrews, in the second millennium BCE. Typically, organized religion celebrated a society's status quo, naturalizing the social order as god-given and unchangeable. The power of organized religion was also manifested in the landscape, often in the form of large religious temples and monuments such as Egyptian or Mayan pyramids or Mesopotamian ziggurats that consumed vast amounts of state resources to construct and maintain. Religion remained by far the dominant ideology in almost all societies, including feudal Europe, until the Enlightenment of the 17th and 18th centuries, when its influence began to wane in the face of growing scientific secularism.

Yet another innovation to emerge from the Neolithic was the use of metals. Preagricultural societies, as we have seen, relied almost exclusively on stone or bone for their tools. The extraction and refinement of metallic ores is a complicated process that requires skill in metallurgy, and was made possible only by the increasingly complex divisions of labor that agriculture facilitated. Metals of different types offered numerous advantages over stone or bone; metal tools and weapons, for example, were stronger, more flexible, and less brittle. The adoption of metalworking is frequently divided into two distinct eras, the Bronze Age and the Iron Age. The Bronze Age, dating roughly between 3000 and 1500 BCE, refers to the discovery of bronze, an alloy of copper and tin. Its appearance varied among different parts of the world, and it probably began first in the Middle East, from where it spread to Europe, Persia, and Egypt, where bronze played a key role in numerous cultures ranging from the Sumerians and Egyptians to early Greece. Similar Bronze Ages can be dated in China, from where it diffused to Japan and Korea, and India. Bronze Age cultures left behind numerous objects such as metal containers, tools, and weapons. The Iron Age, lasting between roughly 1500 BCE and CE 500 (roughly the end of the Roman Empire), saw iron gradually displace bronze as the preferred metal of choice. The Iron Age was most pronounced in Europe, although similar events occurred in China, India, and Africa. Some Iron Age cultures could make steel, which is iron with large quantities of carbon added to give it strength and flexibility, but steel remained a rare good until the Industrial Revolution of the late 18th and 19th centuries.

2.2 The Urban Dawn and Rise of Civilizations

As various preindustrial agricultural systems became widespread over large parts of the world, and with them complex, class-based divisions of labor, the social surplus of such societies grew accordingly, and with it, the ability to sustain a growing number of nonagricultural workers. The interactions among agriculture, the surplus, class, and the state were all fundamental to a phenomenon closely associated with the Neolithic Revolution, the rise of cities. Small villages, of course, had existed in many regions for thousands of years, and continued to be the norm for the majority of the world's people until quite recently. But the growth of large, densely populated societies that could feed themselves regularly and predictably gave rise to the world's first true cities, an event that dramatically changed the economic and cultural landscapes of the Bronze and Iron Age worlds. In short, true cities are impossible without an agricultural surplus, for urbanites rely on the wealth generated by peasants and farmers in the fields. But cities came into their own as distinct centers of power, ideology, culture, and innovation, in turn changing the peoples and places that surrounded them.

Whatever the specific reasons motivating them, early civilizations represented major cultural advances and brought many changes. Although each one had its own distinctive character, they shared several features. Agriculture and urbanization are the two most obvious developments that have impacts on land, but civilization also involves changes in nonmaterial culture: the rise of powerful elites, labor that is submissive to the elite, the disappearance of the egalitarian family-based group, the rise of armies and the technology of war, the growth of bureaucratic systems, opportunities for patronage for artists and scholars, and the institution of private landownership. In many cases religion was the basis of early power, often of a tyrannical system, but all civilizations became increasingly secular.

The city–civilization link is an ancient one; the two words share the Latin root *civitas*. Civilizations create cities, but cities mold civilizations. The earliest cities probably date from about 3500 BCE and developed out of large agricultural villages. It is no accident that the emergence of cities coincided with major cultural advances, such as the invention of writing.

The origins of urbanization and urbanism—the urban way of life—are debatable. It seems likely that most cities originated in one of four ways. First, we know that cities were initially established in agricultural regions, and that city life did not become possible until the progress of agriculture freed some group members from the need to produce food. Thus, in some areas the first cities probably reflected the production of an *agricultural surplus*, possibly as a result of irrigation schemes. Çatal Huyuk in Turkey, and Jericho in Israel/Palestine, reflect this process. Other early cities of this type were located in the Tigris–Euphrates region (present-day Iraq), the Nile Valley (Egypt), the Indus Valley (Pakistan), the Huang Valley (China), Mexico, and Peru, and were associated with the rise of agriculture in these regions. These cities were the residential areas for those not directly involved in agriculture. They were small in both population and area. Populations generally ranged from 2000 to 20,000, although Ur on the Euphrates and Thebes on the Nile may have numbered as many as 200,000. In 2001 a significant discovery was made on the coast of Peru. The city of Caral, located about 200 km north of Lima, has been dated as early as 2627 BCE—about 1000 years before any other city in the Americas. The inhabitants of Caral appear to have supported themselves by practicing irrigation, growing squash, beans, and cotton.

Other roles that cities played were as *marketplaces* for the exchange of local products. Some early cities were located on navigable waterways, others on long-distance trading routes: Venice was a prime example of a port city, while Baghdad was a port in the desert. Many cities may have started as military, defense, or administrative centers. Yet others were important ceremonial centers for religious activity. In short, the rise of cities does not reflect any single economic, cultural, or political imperative, as most cities were significant in all of these domains.

It is important to understand early urban growth in the context of the unique geographic and historical circumstances in which cities began. Cities began in a limited number of places in the world, all of which had previously established productive agricultural systems (Table 2.1).

2.2.1 Mesopotamia

Mesopotamia, Greek for "land between the rivers,"—the Tigris and Euphrates (Fig. 2.1)—is widely held to be the world's first major center of early urban growth. This region is often included as part of the Fertile Crescent, the band of territory that stretches toward the west to include contemporary Syria, Lebanon, and Israel. Many different cultures and civilizations

Table 2.1 Chronology of some early civilizations.

Egypt	3000–332 BCE
Minoan	3000–1450 BCE
China	3000 BCE–present
Andean	3000–1800 BCE
Indus	2500–1500 BCE
Mesopotamia	2350–700 BCE
Mycenaean	1580–1120 BCE
Olmec	1500–400 BCE
Greece	1100–150 BCE
Rome	750 BCE–CE 375
Monte Alban	200 BCE–CE 800
Moche and Nazca	200 BCE–CE 700
Teotihuacan	CE 100–700
Maya	CE 300–1440
Tiahuanaca	CE 600–1000
Toltec	CE 900–1150
Chimo and Inca	CE 1100–1535
Aztec	CE 1200–1521
Benin	CE 1250–1700

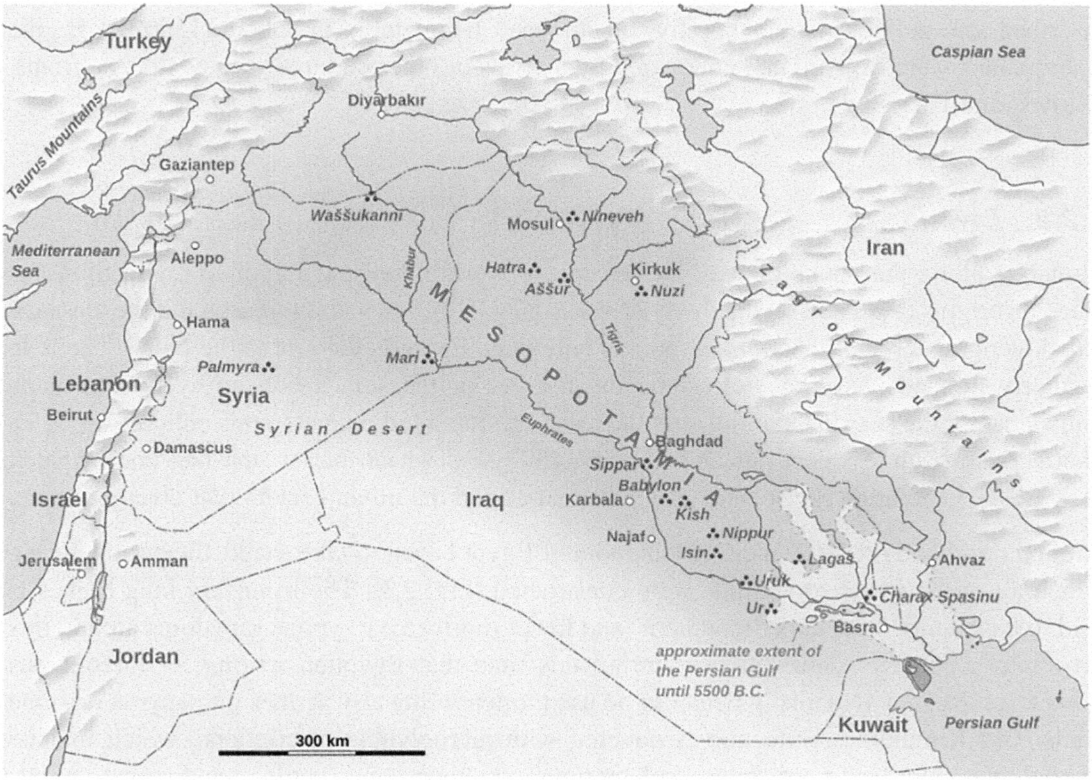

Figure 2.1

Mesopotamia is the region where the world's first cities began.
(https://commons.wikimedia.org/wiki/File:N-Mesopotamia_and_Syria_english.svg.)

occupied this area over time. With the domestication of barley and wheat, the first Mesopotamian culture to arise was the Sumerians, around roughly 3500 BCE. They built cities such as Ur, Uruk, Eridu, and Lagash, all noted in the Old Testament, which were likely the first cities in the world. The wheel appeared here around the same time. Evidence of the powerful role of organized religion was to be seen in the ziggurats, or stepped pyramids that the Sumerians constructed. They also invented an early system of writing, or cuneiform, much of which consisted of symbols pressed into clay. The Sumerians were the first to divide the year into 12 months, the month into four weeks, the day into 24 hours, and the hour into 60 minutes; every time you look at a clock or watch you are indebted to them. The Sumerians were displaced in due course by other early cultures such as the Babylonians (roughly 1900 BCE), who constructed the famed city of Babylon, the world's largest metropolis at the time. Under their famed king Hammurabi, Babylonian law was codified and standardized, and inscribed on public monuments for all to see. Ultimately, by about 1500 BCE, the Babylonians were overrun and conquered by a series of other cultures such as the Assyrians, Hittites, and Persians.

2.2.2 Egypt

Stretched along the banks of the Nile River, the world's longest, Egyptian civilization has since its origins depended on that body of water (Fig. 2.2); not without reason did the classical Greek geographer and historian Herodotus note that "Egypt is the gift of the Nile." The Nile regularly flooded its valley, refreshing the fertility of the soils there, an event that made knowledge of the seasons, rainfall, and the river's rise and fall crucial to this culture. Along its banks toiled countless peasants and farmers, who grew wheat, barley and flax and irrigated their fields, generating the social surplus that sustained the urban centers that emerged there.

Around roughly 3100 BCE the Old Kingdom of Egypt began to take shape, the period during which all of the famous pyramids were constructed (Fig. 2.3). The legendary king Menes is said to have unified the upper (southern) and lower (northern) Egyptian kingdoms around this time into a coherent entity. It was during this time that Egyptian writing, or hieroglyphs (Greek for "sacred symbols"), began to be used, often written by scribes on papyrus or stone walls. Old Kingdom architecture is covered with hieroglyphic inscriptions, which include mythologies and religious tales, celebrations of kings and battles, and royal edicts. Hieroglyphs were indecipherable after the collapse of ancient Egyptian culture, but were eventually translated by French linguist Louis Champollion after the discovery of the Rosetta Stone in 1799, which contained similar inscriptions in three different languages; soon thereafter, modern Egyptology was born.

Like all classical cultures, ancient Egyptian culture was saturated in religious cosmologies that painted detailed portraits of its many gods and goddesses. Ra, the sun god, was the most important of these, traveling across the sky by day, dying in the west at night, only to be reborn in the east the next morning. Other deities included Nut the sky goddess, Osiris and Isis, and Horus.

At the apex of ancient Egyptian culture was the pharaoh, a king who assumed divine properties (revealing the role of religion in naturalizing social events in these cultures). Pharaohs of varying political statures founded the dynasties that characterized ancient Egyptian history. Their burial tombs, the pyramids, some of the most impressive structures of the ancient world, concentrated vast quantities of state resources into specific sites; they were designed to protect the pharaoh on his voyage through eternity (as was his body through mummification), and ranged in size and height depending on the wealth of their builders. Roughly 60 were built during the heyday of the Old Kingdom. Estimates of the labor required for their construction range between 10,000 and 20,000 men working for 20 years. Most are

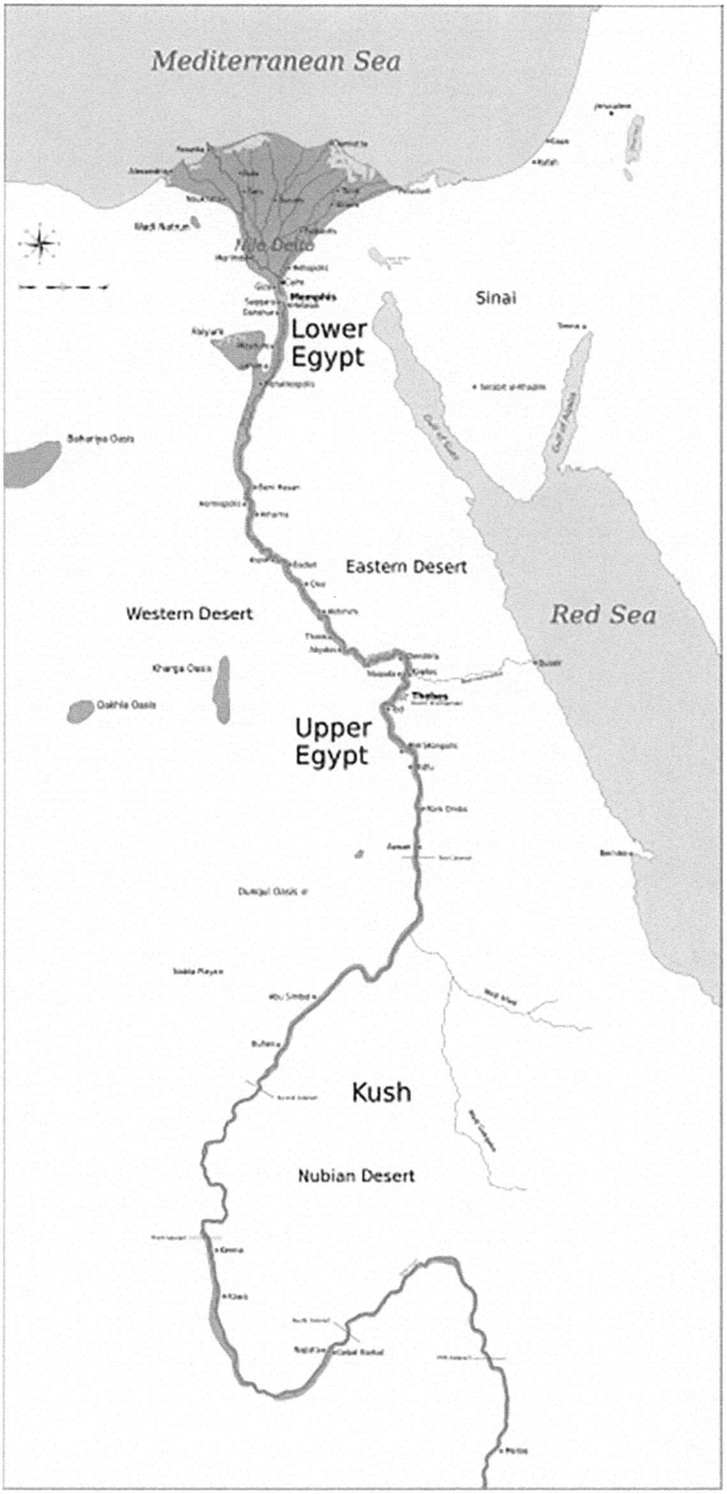

Figure 2.2

Egypt's civilization was essentially a ribbon on the Nile River.
Map lines delineate study areas and do not necessarily depict accepted national boundaries.
(https://commons.wikimedia.org/wiki/File:Ancient_Egypt_map-en.svg.)

Figure 2.3
Egypt's pyramids were elaborate tombs for pharaohs and the wealthy.
(https://commons.wikimedia.org/wiki/File:The_Pyramids_of_Giza_in_Egypt.jpg.)

concentrated in northern Egypt, particularly on the outskirts of modern Cairo, including the Giza complex, site of the Great Pyramid, the Chefren pyramid, and the Sphinx.

The New Kingdom, which eventually displaced the Old Kingdom, witnessed the locus of power and construction shift to the south of the country, such as the famous site of Luxor. The most famous of the New Kingdom pharaohs, Ramses II, led enormous military campaigns to subdue the Kushitic kingdoms in the south and northwards into what is now Palestine and Lebanon. A great builder, Ramses also constructed the enormous statues of pharaohs at Abu Simbel, originally located on the banks of the Nile but subsequently moved in the 1970s to protect it from flooding.

Ancient Egyptian society eventually collapsed by the third century BCE, when it was overrun by a series of foreign powers. Alexander the Great made the region part of his empire, establishing the city of Alexandria at the mouth of the Nile as his capital. Alexandria became an important port city and center of learning for centuries thereafter. The Romans conquered Egypt shortly before the time of Christ, when Augustus Caesar made it his personal domain and the breadbasket of the empire. Later, in the seventh century, the Arabs overran the country, founding the city of Cairo and giving it its current Islamic composition.

2.2.3 China

A third great center of early civilization was China, the oldest continuously inhabited culture on the face of the Earth. China had been populated for millennia during the Paleolithic, but classical Chinese culture began to take shape around 3500 BCE in the Huang He river valley in northern China (Fig. 2.4). Wheat and barley were the staple crops cultivated here; only later, in southern China, did rice cultivation become widespread, giving rise to the enormous number of paddies that cover the landscapes of the southern part of the country.

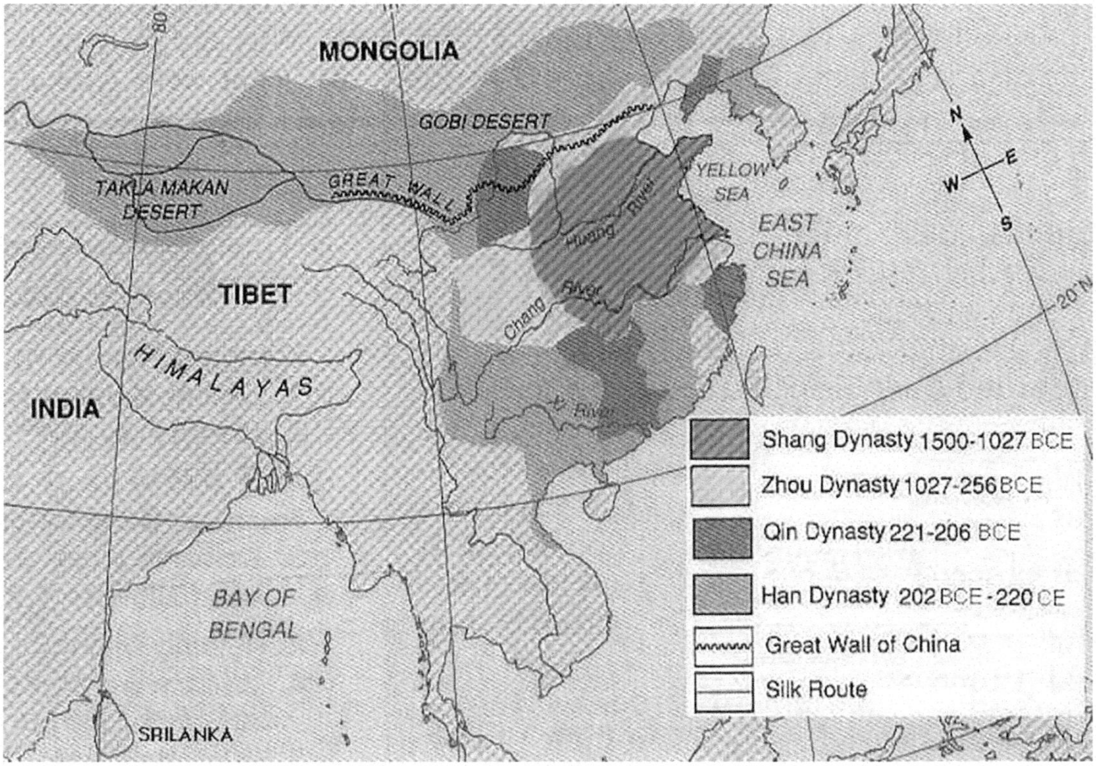

Figure 2.4
China's first cities arose in the north along the banks of the Huang He River.
Map lines delineate study areas and do not necessarily depict accepted national boundaries.
(https://gohighbrow.com/early-chinese-civilization/.)

Bronze Age China was defined politically by the Shang dynasty, China's earliest (1500 BCE), which expanded to cover the central part of the nation. Some of the country's earliest urban centers, such as Anyang, arose during this era. Early Chinese history was often marked by squabbling kingdoms. In 221 BCE, the famous emperor Shih Huang-ti unified the country, founding the Chin dynasty (from which the word "China" is derived). It was around this time that Confucianism became institutionalized as state ideology, naturalizing the social order and proscribing strict roles for every member to follow. The Chin also started the Great Wall to protect themselves from raids by the pesky Mongols to the north. One of ancient China's most memorable sites is the former capital of Xian, where emperors built the famous terra cotta army of soldiers, each individually designed, to accompany them in the afterlife. Other early dynasties of note included the Han, roughly contemporaneous with the Roman Empire, which expanded Chinese control westwards along the important Silk Road routes (Fig. 2.5).

Classical China was famous as a source of innovation. Indeed, China was, on average, much larger and wealthier than Europe for most of its history; not until the European conquests that began in the 18th century did China fall behind. Its stream of inventions raised productivity levels to unparalleled heights. For example, the Chinese invented papermaking and block

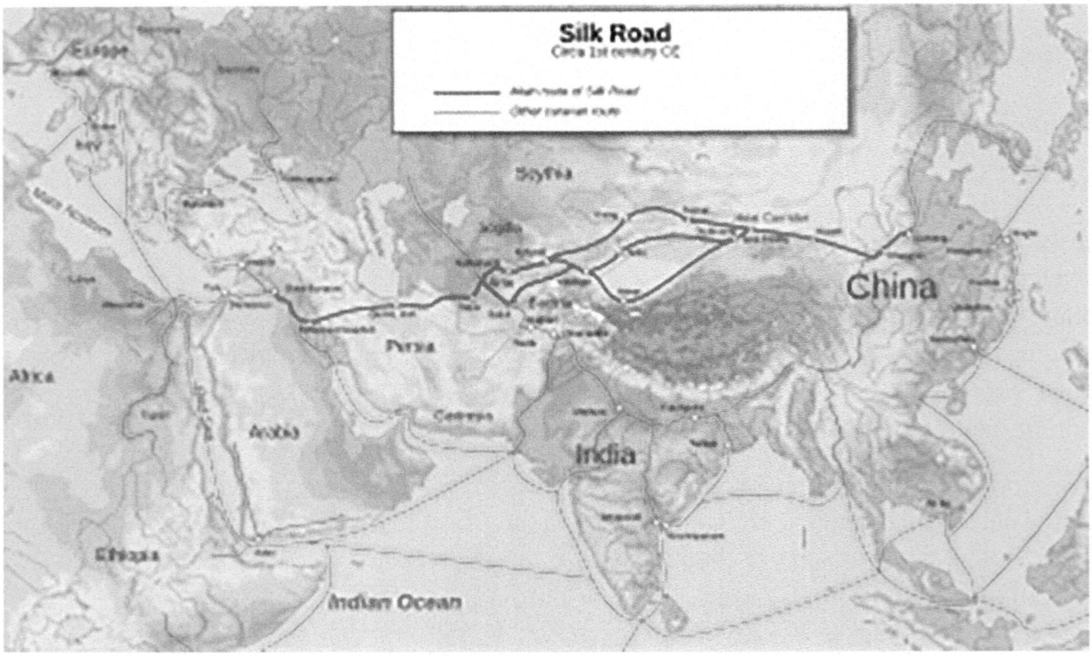

Figure 2.5
The Silk Road was an important system of trade that connected China and the West for 2000 years. Map lines delineate study areas and do not necessarily depict accepted national boundaries. *(https://commons.wikimedia.org/wiki/File:Silk_road_with_hexi_corridor_labelled,_1st_century_ce.svg.)*

printing; the Arabs, capturing Chinese soldiers in a war, learned of these innovations and eventually transmitted them to Europe. Other Chinese inventions included the field pump (to draw water up from rivers to irrigate crops), the crossbow, the iron plow, the umbrella, wheelbarrow, movable type, paper money, the stirrup, windmill, and gunpowder (this was created to frighten Mongol horses with firecrackers during their attacks). Most of these phenomena found their way westward along the Silk Road, gradually percolating into medieval Europe and helping to reshape that society.

2.2.4 The Indus River Valley

Along the banks of the mighty Indus River, in what is now Pakistan (Figure 2.6), a fourth major early center of civilization arose around 2600 BCE, only to collapse suddenly around 1700 BCE. The people who populated this region were likely the Dravidians, an ethnic and linguistic family of languages that is found today primarily in southern India, considerably far to the south of the Indus.

The Dravidians created a number of flourishing urban centers along the Indus, of which the two most famous were Harappa and Mohenjo-Daro. Each likely contained tens of thousands of people, fed by vast networks of wheat fields nearby. Because the script of this culture has

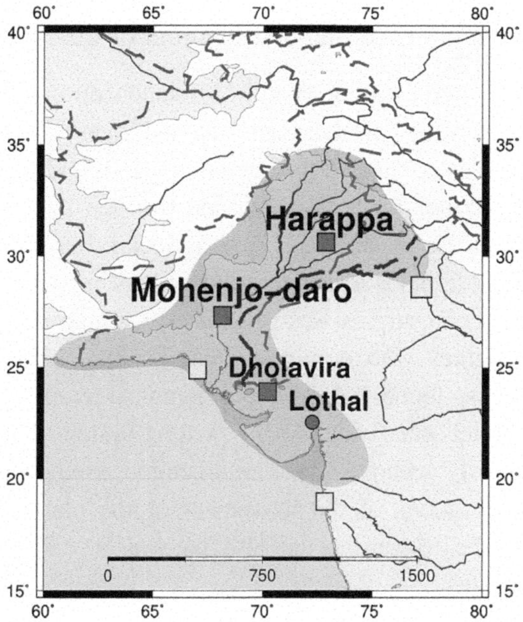

Figure 2.6
The Indus River, in what is today Pakistan, was an early source of urbanization in South Asia. Map lines delineate study areas and do not necessarily depict accepted national boundaries.
(https://commons.wikimedia.org/wiki/File:Simple_map_of_Indus_valley_civilization.svg.)

yet to be deciphered, much remains unknown about them. Clearly they had extensive trading ties with India, Central Asia, and the Middle Eastern world, exchanging goods by caravan and by sea routes. They seem to have had an elaborate calendar and knowledge of advanced mathematics, including prime numbers.

The collapse and extermination of the Indus River valley culture around 1700 BCE has been the subject of considerable speculation. Some hold that it fell due to prolonged drought and ensuing famine. But the most persistent explanation is that the Indus River Dravidians fell victim to invaders from the northwest, the Indo-Europeans, who had domesticated the horse and went on to conquer the northern half of the Indian subcontinent, driving the Dravidians to the southern part of the peninsula, where they remain today.

2.2.5 Mesoamerica

Mesoamerica, the isthmus that includes Mexico and Central America, was also an important early center of urbanism and civilization. Numerous cultures occupied this region at different moments in time. Around 1500 BCE, the Olmecs started the first major society in this area, building enormous stone heads, the purpose of which has never been discovered. Between CE 200 and 700, the city of Teotihuacan, located near what is now Mexico City, established control over a vast region of southern Mexico; in the city's borders, its residents built giant pyramids known as the Pyramid of the Sun and Pyramid of the Moon (Fig. 2.7).

Perhaps the best known and longest lasting of the Mesoamerican cultures were the Maya, who inhabited a series of city-states in the Yucatan Peninsula and what is today Belize and Guatemala (Fig. 2.8). Dozens of Mayan cities dominated the coastal lowlands and the highlands of the interior, including, to name a few, Calakmul, Caracol, Tikal, and Chichen Itza. Mayan societies sustained themselves through widespread use of slash-and-burn agriculture or shifting cultivation, growing their staple crop, maize or corn, in fields irrigated through a vast network of canals. It was through this process that much of the social surplus was generated to sustain the Mayan elite, the cadre of kings, priests, and military figures who dominated these cultures. The Maya built a series of pyramids, many larger than those found in Egypt, as ceremonial temples (Fig. 2.9). Mayan culture also exhibited a profoundly intellectual side, with a writing system that combined phonetic and pictographic symbols, a famously accurate calendar structured around multiple cycles of creation and destruction, detailed observations of the movement of the planets (especially Venus), and knowledge of mathematics. The Mayans could also be violent, and the city-states engaged in frequent warfare with one another.

The collapse of Mayan society around CE 900 has been hotly debated. Some maintain that it was due to prolonged droughts that diminished the food supply. Others hold that over-cultivation of the soils through swidden agriculture contributed to loss of vegetation cover, soil erosion, and declining crop fertility. Yet other explanations point to foreign invaders. It is

Figure 2.7
Teotihuacan, including the Pyramids of the Sun and Moon.
(https://commons.wikimedia.org/wiki/File:Vista_a%C3%A9rea_de_Teotihuacan_2024_01.jpg.)

clear that by CE 1000, the civilization was over, although southern Mexico and Guatemala continue to have large Mayan populations today.

The successors of the Mayans were the Aztecs, who overran a large swath of southern Mexico (Fig. 2.10), establishing an important but short-lived empire in the 14th century that vanished during the Spanish colonial onslaught of the 16th century. Unlike the Maya, who formed a network of independent city-states, the Aztecs concentrated power in a single capital city, Tenochtitlan, on the site of what is now Mexico City. According to Aztec mythology, the city was founded on a site where an eagle was observed eating a snake, a motif that appears on the Mexican flag today. From their capital, the Aztecs established military authority over numerous surrounding tribes, many of which resented their control. They also supervised extensive trade routes in the region that exchanged goods such as jade, obsidian, seashells, parrot feathers, and precious metals.

The Aztecs are often viewed as a particularly militaristic culture, in which the primary function of the state was waging war, often for the purposes of capturing slaves and prisoners for sacrifice. Thus, the empire wielded a large army, including an elite guard (the Mexica, from which comes the label "Mexico"). Prisoners were often sacrificed in large numbers during important religious occasions, their blood used to quench the thirst of the sun god, Huitzilopochtl. Typically this event saw priests cut the heart out of the living victim, and the body was tumbled down the steps to the masses below.

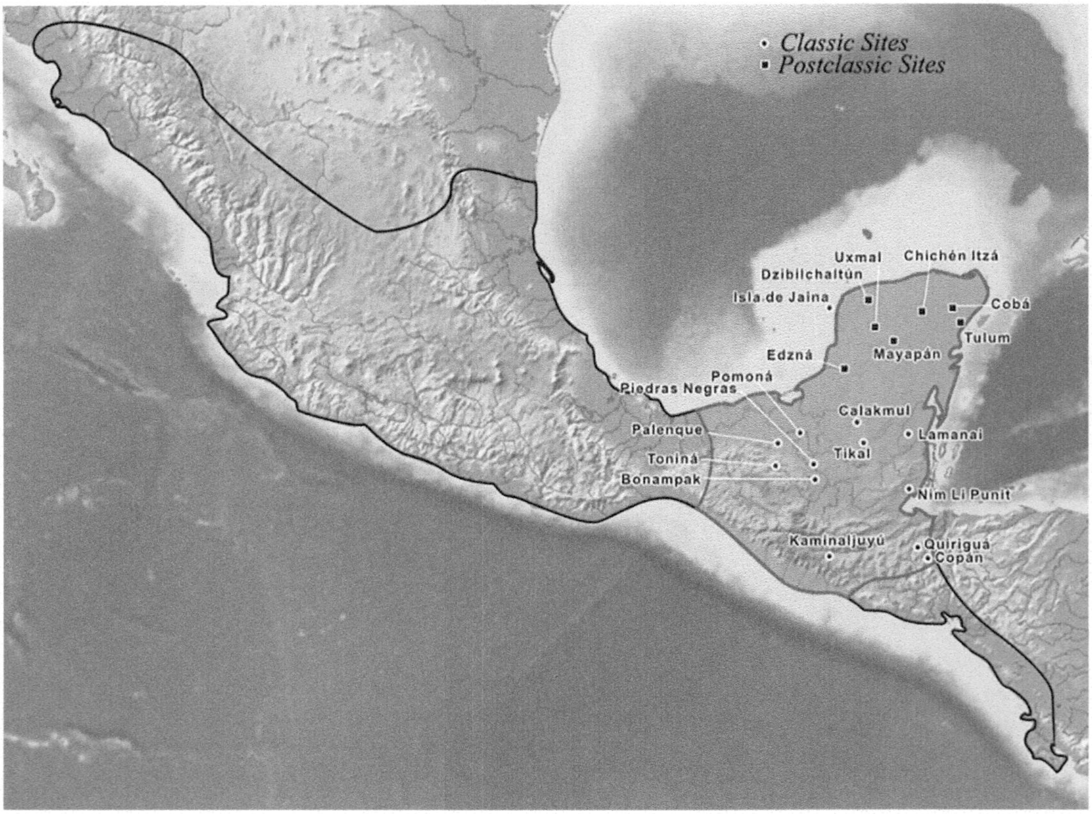

Figure 2.8

The Mayan realm, in southern Mexico, Belize, and Guatemala, was a major region of early urbanization in the Western Hemisphere.
Map lines delineate study areas and do not necessarily depict accepted national boundaries.
(https://commons.wikimedia.org/wiki/File:Mayamap.png.)

In the 1520s, the Spanish conquistador Hernando Cortes famously led a small group of soldiers into Tenochtitlan and overthrew the Aztec emperor, Moctezuma (Fig. 2.11). How did such a tiny group of foreigners manage such a feat? In *Guns, Germs, and Steel*, Jared Diamond argues that the Spanish had numerous technological advantages, including horses, iron swords and shields, guns, and attack dogs. Others note that tribes conquered by the Aztecs assisted the Spanish. In the wake of the European conquest, smallpox spread rapidly throughout the Americas, effectively depopulating much of Mexico as the Indigenous peoples had no biological resistance to this foreign virus.

2.2.6 South America

Between the Andes Mountains and the western shore of South America, a variety of cultures arose, gave birth to civilizations, and eventually fell. Most of these cultures relied heavily on

Figure 2.9
Mayan pyramids, such as this one at Chichen Itza, were as large as those in Egypt.
(https://commons.wikimedia.org/wiki/File:Chichen_Itza,_El_Castillo_(14180679857).jpg.)

the potato, which they domesticated and grew in 700 different types adapted to a variety of microclimates at different elevations. Chavin culture, for example, arose around 1700 BCE in coastal Peru, supplementing potatoes with seafood, notably anchovies. The Nazca people drew elaborate chalk designs of surprisingly large size and accuracy on the desert landscapes. The Moche people built large temples and practiced human sacrifice.

But it was the Inca who created the largest empire of South America, stretching along the coast for almost 2,000 miles (Fig. 2.12) from the jungles of Colombia to the Atacama Desert. In the Quechua language that they spoke, the empire was called Tawantinsuyu ("land of the four quarters"). Like the Aztecs in Mexico, the Inca were a tribe that conquered numerous neighboring peoples in an empire that lasted only about two centuries. From their capital city of Cuzco high in the Andes, the Inca administered an empire that was connected by a vast network of roads (Fig. 2.13), along which traveled messengers on foot, llama and alpaca caravans carrying foodstuffs, slaves, and soldiers. The Inca had no wheel or system of writing, but did develop a counting system using quipu, or woven cloth with a systematic pattern of knots. By appealing to the sun god, Inti, they elevated the power of the leader, the Sapa Inca, into the status of a demi-god.

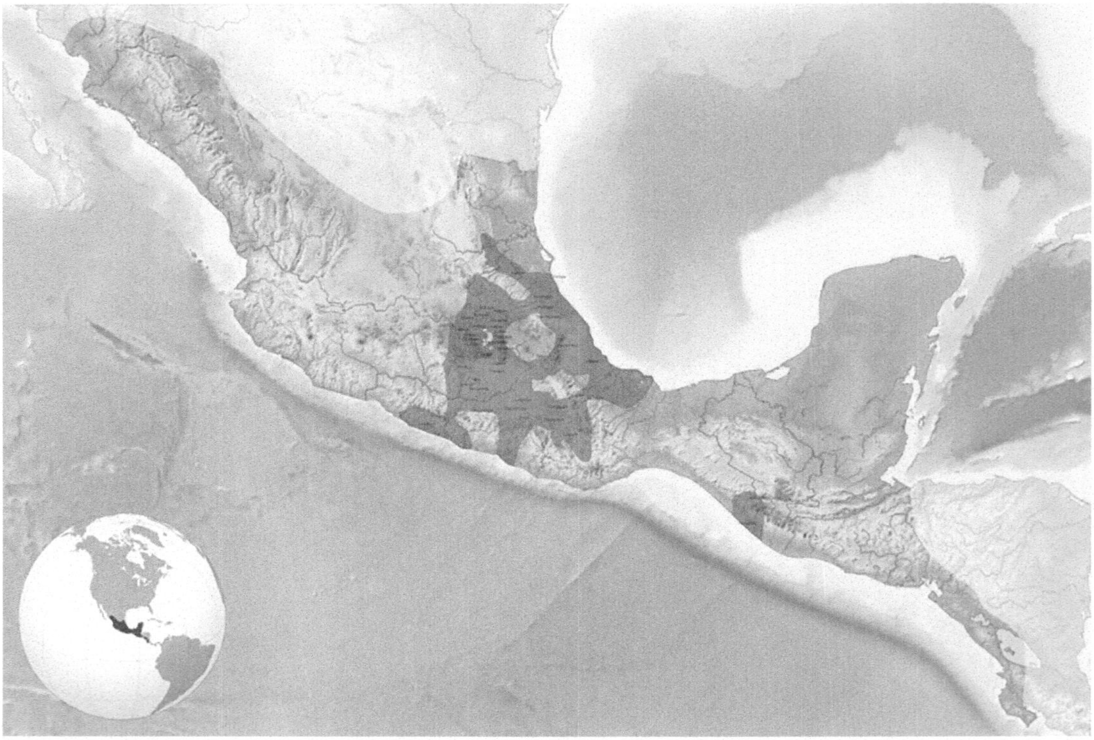

Figure 2.10
The Aztecs dominated central Mexico, and their capital, Tenochtitlan, later became Mexico City. Map lines delineate study areas and do not necessarily depict accepted national boundaries. *(https://commons.wikimedia.org/wiki/File:Aztec_Empire.png.)*

The Inca met a fate essentially identical to that of the Aztecs further north. In 1526 the Spanish conquistador Francisco Pizarro led a small group of men and captured the last Sapa Inca, Atahualpa. After holding him for ransom, they executed him. Deprived of their leader, the empire began to crumble rapidly, assisted by the Spanish advantage in military technology. Smallpox, too, took a horrendous toll and exterminated millions of people. The Spanish enslaved many survivors, working them to death in the silver mines of the Andes, such as the city of Potosi, the largest silver-producing location on the planet.

In 1911, the last Incan city, Machu Picchu, was discovered high in the Andes (Fig. 2.14). Far from the area of Spanish control, it escaped the destruction that other Incan cities met, and remains an important monument to this day to the power and sophistication of this civilization.

2.2.7 North America

Many earlier interpretations of native cultures in North America erroneously portrayed them as populated only by hunters and gatherers. In fact, long before Columbus, several centers of

Figure 2.11
This painting depicts June 30, 1520, when Hernando Cortés and his men decapitated the Aztec Empire in what is known as la noche triste, the Night of Sorrows.
(https://commons.wikimedia.org/wiki/File:Conquista-de-M%C3%A9xico-por-Cort%C3%A9s-Tenochtitlan-Painting.png.)

urbanization existed in what is now the United States. The largest of these is loosely referred to as the Mississippian mound building culture, after the groups of tribes who built thousands of burial mounds of varying sizes over a large area extending from Minnesota to Ohio to northern Florida. Many of these peoples relied on corn as their staple crop, which had diffused northwards from its site of domestication in Mexico. The largest pre-Columbian North American urban center was Cahokia (Fig. 2.15), a city that flourished between CE 900 and 1200 near what is now the city of St. Louis. At its peak, Cahokia may have had 12,000 residents, and its remnants today are home to some of the largest mounds built by that culture.

Far to the southwest, in the Sonoran desert of Arizona and New Mexico, a series of small urban centers arose populated by the Anasazi people between CE 450 and 1350 (Fig. 2.16). In an unforgivingly dry and hot climate far from major rivers, they constructed a series of impressive villages nestled at the base of enormous cliffs (Fig. 2.16) replete with multistory buildings and sweat lodges. Elaborate pottery and weaving testify to their skills. Likely due to prolonged drought, or perhaps invasion from Mexico, the Anasazi people vanished in the 14th century; their descendants are the Hopi and Navajo today.

Figure 2.12
The Inca Empire dominated the western coast of South America.
Map lines delineate study areas and do not necessarily depict accepted national boundaries.
(https://commons.wikimedia.org/wiki/File:Tawantinsuyu_(orthographic_projection).svg.)

2.2.8 Africa

In large part due to centuries of ethnocentric Western history, sub-Saharan Africa has often been overlooked as a center of early civilization. While Egypt is often seen as the most elaborate early African society, several parts of Africa south of the Sahara also established flourishing kingdoms (Fig. 2.17). In Western Africa, for example, along the Niger River, the Mali, Ghana, and Songhai kingdoms played important roles in organizing vast numbers of people, growing a variety of staple crops (e.g., yams), and building urban centers.

Many of these kingdoms became wealthy by taxing the trading caravans that crossed the Sahara Desert using camels, connecting the Arabs to the north with the dark-skinned peoples of the coastal areas (Fig. 2.18). From the north the Arabs provided salt; from the south, Africans provided gold, ivory, textiles, and slaves. It was along these routes that Islam also began to enter into Western Africa, where it plays an important role today. The most famous of the Western African cities was Timbuktu, located at the great bend of the Niger, which became important not only for its commercial success but as an intellectual center as well, with numerous universities. When the Europeans began to set up maritime trade routes in the 16th century, they effectively put an end to the trans-Saharan caravans, and the great cities of the interior declined in importance and wealth.

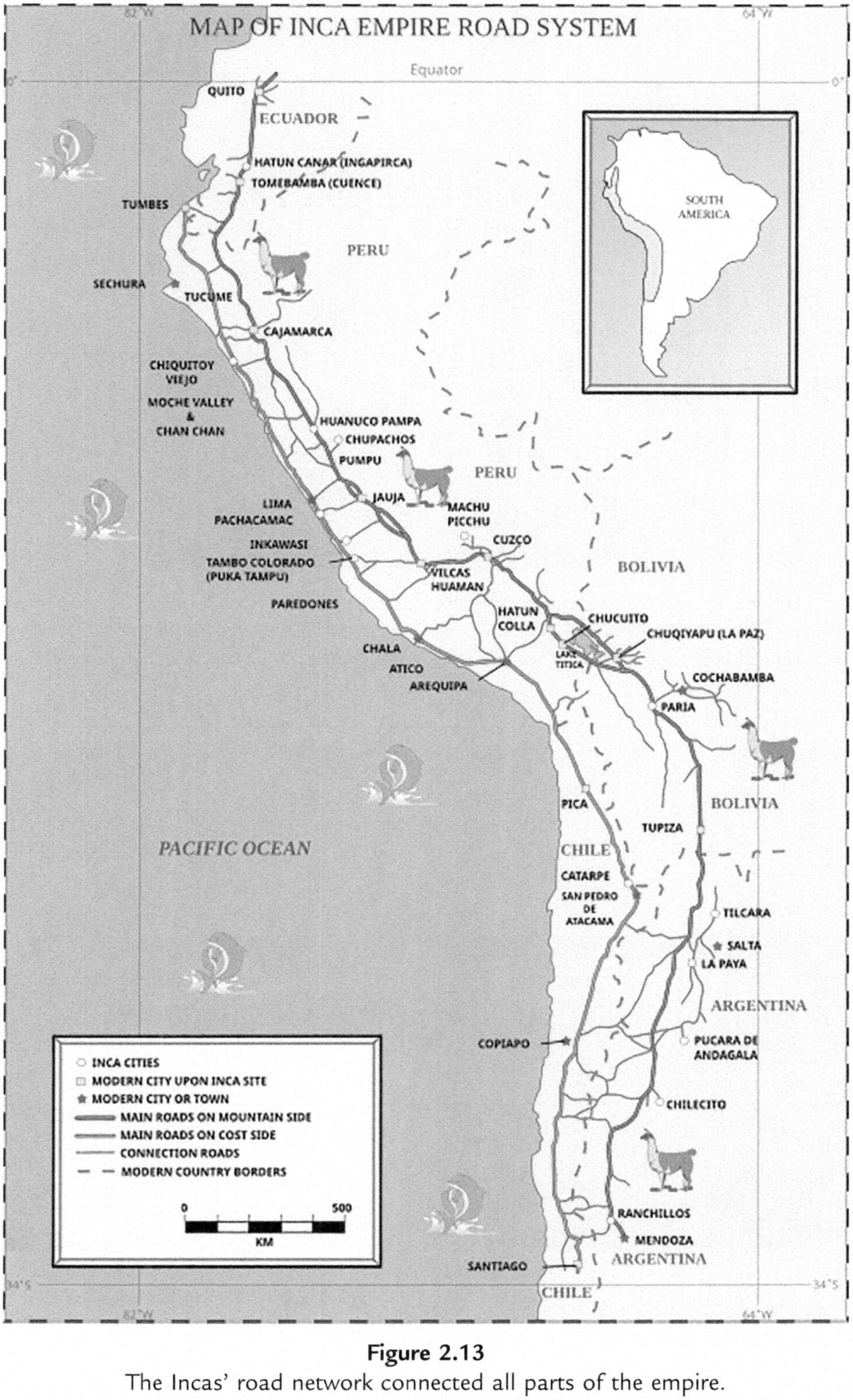

Figure 2.13

The Incas' road network connected all parts of the empire.

Map lines delineate study areas and do not necessarily depict accepted national boundaries.

(https://commons.wikimedia.org/wiki/File:Inca_road_system_map-en.svg.)

Figure 2.14
Machu Picchu, the so-called hidden city of the Andes, was rediscovered in 1911.
(https://commons.wikimedia.org/wiki/File:80_-_Machu_Picchu_-_Juin_2009_-_edit.jpg.)

Figure 2.15
Artist's representation of Cahokia.
(https://commons.wikimedia.org/wiki/File:Cahokia_Mounds_–_UNESCO_reconstruction.jpg.)

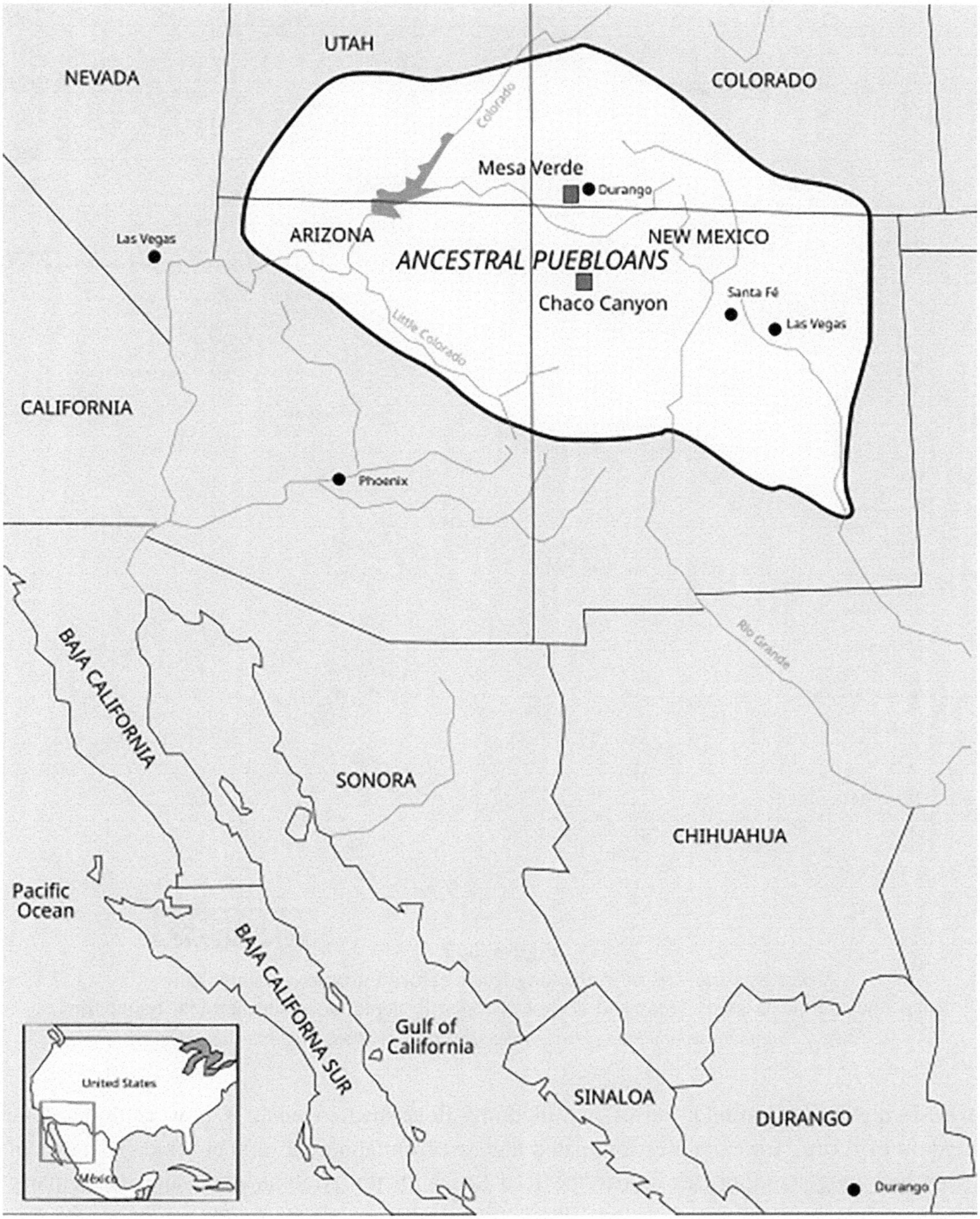

Figure 2.16

The Anasazi peoples of the southwestern United States created towns in a formidable desert environment.

Map lines delineate study areas and do not necessarily depict accepted national boundaries.

(https://commons.wikimedia.org/wiki/File:Ancestral_Puebloan_territory.svg.)

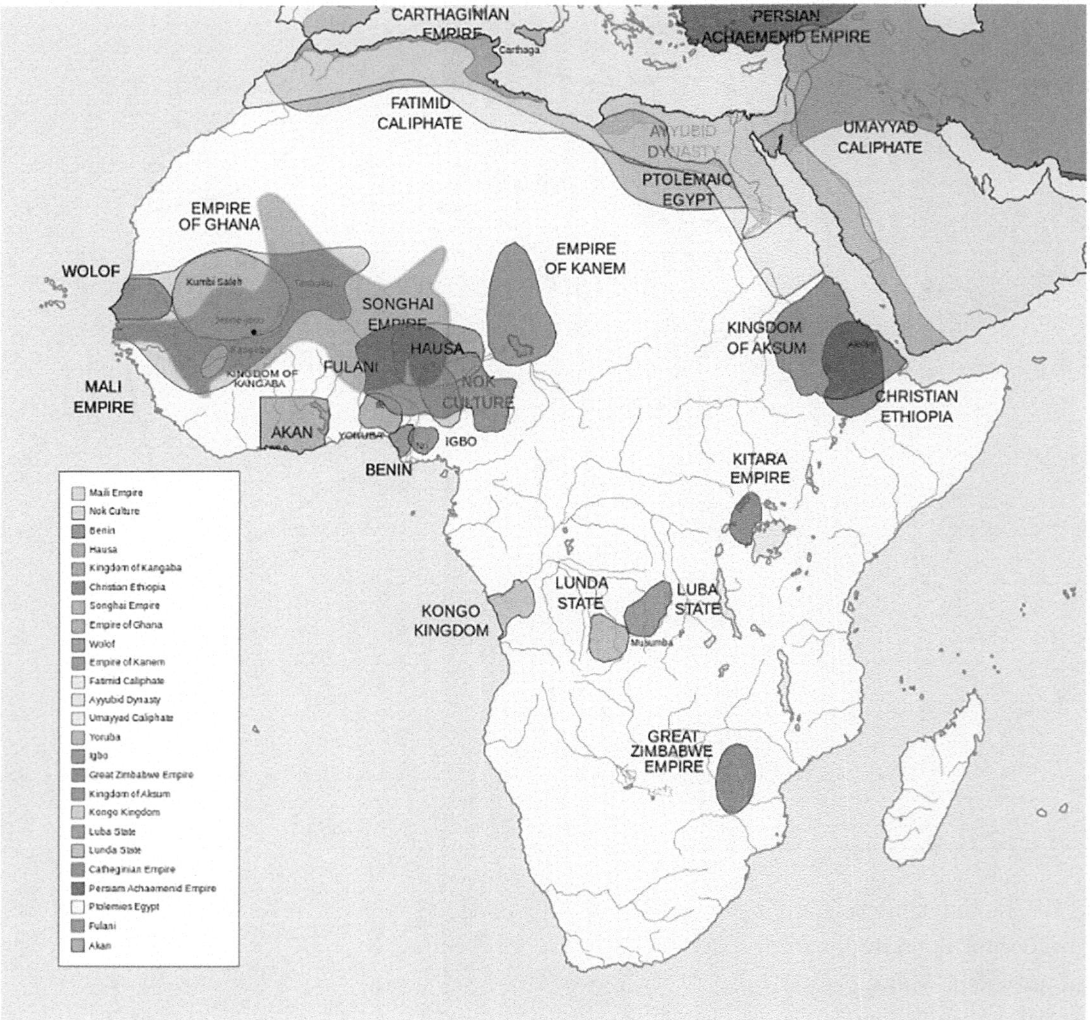

Figure 2.17

Africa was the site of many kingdoms before European colonialism.
Map lines delineate study areas and do not necessarily depict accepted national boundaries.
(https://commons.wikimedia.org/wiki/File:African-civilizations-map-pre-colonial.svg.)

In Eastern Africa, a similar series of kingdoms flourished in what is now Ethiopia. The kingdom of Axum, for example, dominated the area from about CE 100 to 1000. This part of Africa had long-standing ties across the Red Sea with the Arabian peninsula, particularly Yemen; it was from Arabia that first Christianity, and later Islam, found their way into this part of the continent. By the 11th century, as Islam shifted the locus of wealth and power to the north, Axum entered into a long period of terminal decline.

Figure 2.18
Ninteenth century painting of the camel caravans approaching Timbuktu.
(https://commons.wikimedia.org/wiki/File:TIMBUKTU-EINZUG.jpg.)

In Southern Africa, the Zimbabwe kingdom developed a series of urban centers in its own right between CE 1100 and 1500 (Fig. 2.19). With exports of slaves, ivory, gold, and cattle hides, the Zimbabwean kings connected this part of the country with Arab traders along the coast; their goods also tied them loosely to even farther centers of wealth across the Indian Ocean in what was a flourishing international network that crossed that body of water for centuries until the European arrival in the 16th century. The modern country of Zimbabwe takes its name from this empire.

2.2.9 Europe

Although it is not necessarily one of the first places in the world to undergo early urbanization, Southern Europe nonetheless also emerged as the site of a series of important and influential civilizations.

In Bronze Age Greece, some of the earliest signs of metalworking and other artifacts of urban culture were found on the Cycladian islands and also on the important isle of Crete, home to the famed Minoan culture (Fig. 2.20). The Minoans were a maritime-oriented society that relied heavily on the seas for sustenance. This civilization flourished by means of its extensive

Figure 2.19
Ruins of a large building in Zimbabwe, the country whose name is derived from a kingdom that existed roughly 1000 years ago.
(https://commons.wikimedia.org/wiki/File:Great-Zimbabwe-ruins-outer-walls-3–1200.jpg.)

trading ties throughout the eastern Mediterranean Sea, notably with the Phoenicians and Egyptians. Maritime trade routes can be deciphered from patterns of shipwrecks, some of which still carry the amphora, or clay vessels used to hold goods such as wine or olive oil. While 19th century historians sought to minimize the influence of Middle Eastern cultures on early Greece in order to enhance the sense of a unique set of European origins, more recent scholarship points to important influences from its neighbors to the south, including pottery and the introduction of the Minoan alphabet, or Linear A. On Crete, the Minoan imperial center, Knossos, was renowned for its elaborate palaces and festivals that included young athletes dancing on the horns of bulls. Minoan culture came to a sudden, violent end in 1635 BCE when an enormous volcanic explosion on the island of Thera likely sent a huge tidal wave over the island (Fig. 2.20), effectively exterminating its population.

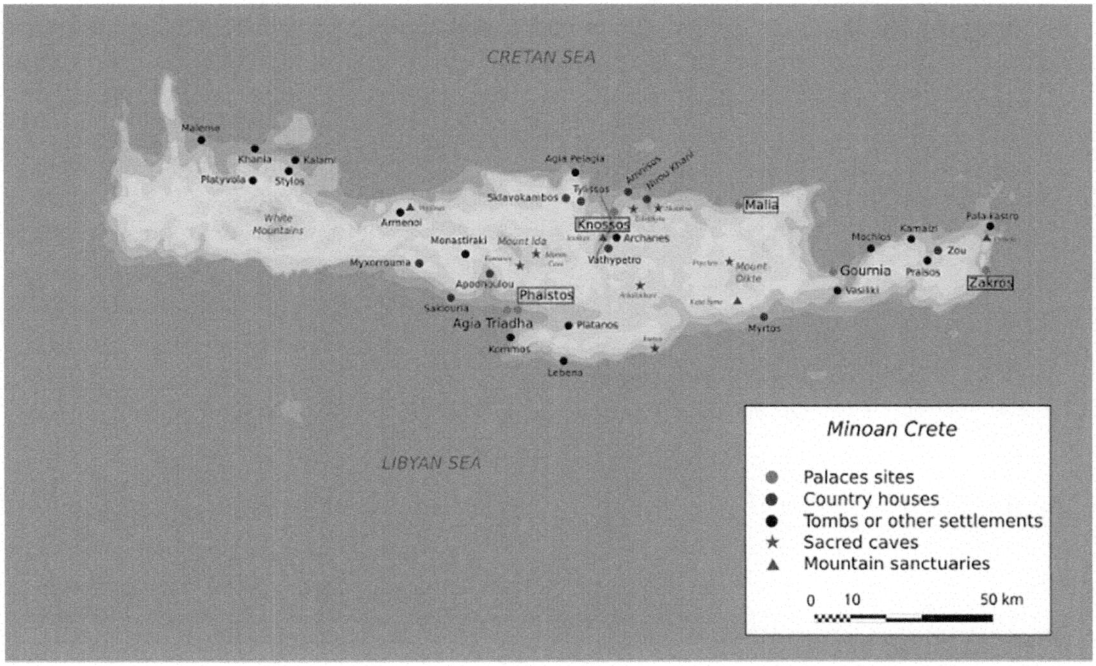

Figure 2.20

Minoan culture was located on the island of Crete, now part of Greece.
M ap lines delineate study areas and do not necessarily depict accepted national boundaries.
(https://commons.wikimedia.org/wiki/File:Map_Minoan_Crete-en.svg.)

On the Greek mainland, Mycenaean culture formed an important nucleus in the series of city-states that later would develop into classical Greece. Located high in the hills, Mycenae was a military fortress that exerted control over a vast part of the Peloponnesian Mountains and coastal areas. The influx of Doric and Ionian Greeks around 1100 BCE, part of the broader Indo-European conquest of Europe, witnessed rival city-states emerge, such as Troy, located near the Dardanelles, or the narrow strait between the Aegean and Black Seas. Troy, of course, was destroyed at the end of the Trojan War, an event portrayed by the blind poet Homer and mythologized in his famous *Iliad*. The ruins of both Mycenae and Troy were discovered by famed German archaeologist Heinrich Schliemann in the late 19th century.

Classical or Hellenic Greece, which flourished between 800 and 400 BCE, witnessed a series of city-states rise in the coastal lowlands not only of Greece, but throughout much of the northern Mediterranean. Greek colonies were established in what is now Spain, southern France (e.g., Marseilles), southern Italy (e.g., Naples) and Sicily, Asia Minor (or Turkey), and the Black Sea. Plato famously likened these to "frogs around a pond" (Fig. 2.21). The city-state, or polis (which is found in words such as Minneapolis or Indianapolis), formed the primary unit of Greek society; not a unified empire, the Greeks, like the Mayans, structured

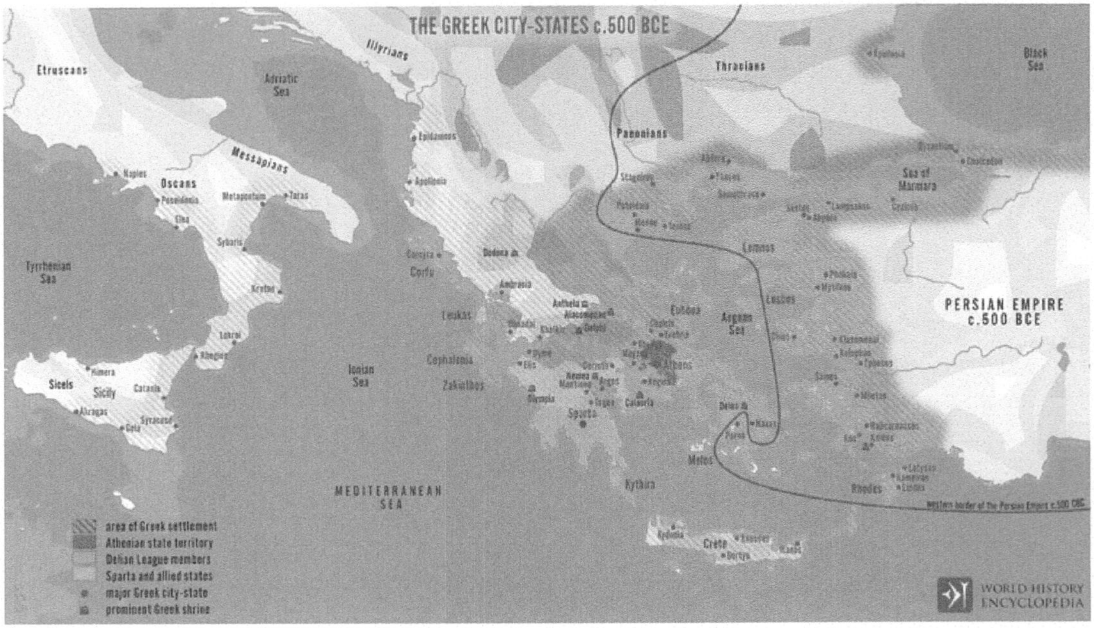

Figure 2.21

The classical Greek world straddled both sides of the Aegean Sea.
Map lines delineate study areas and do not necessarily depict accepted national boundaries.
(https://www.worldhistory.org/image/14624/the-greek-city-states-c-500-bce/.)

their world around independent entities that traded and often feuded with one another. The largest of the polis was Athens, particularly during its zenith in the 5th century BCE, when it became a major center of learning, science, philosophy, art, and scholarship, with famous buildings such as the Parthenon, located in the Acropolis (Fig. 2.22). In addition to wars among themselves (e.g., the Peloponnesian War between Athens and Sparta), the Greeks faced frequent invasions and threats of conquest by the Persian empire to the east.

In the 5th century BCE, Greece fell to its larger, far more powerful neighbor to the west, the Romans. The Roman Republic, and later Empire, which lasted from roughly 700 BCE to CE 476, represented the largest social formation before the Industrial Revolution and thus was the apex of the slave-based mode of production. At its peak in the 2nd century AD, it stretched from Scotland in the northwest to encompass much of Western Europe, all of the Mediterranean Sea and North Africa, and parts of the Middle East such as Palestine and Iraq (2.23). The system was connected by extensive maritime trade routes as well as land-based routes that crossed the large network of well-built roads that the Romans constructed (Fig. 2.23). During the famed Pax Romana, or Roman peace, the region experienced rising prosperity for several centuries. Throughout the empire, the Romans built roads, bridges,

Figure 2.22
The Parthenon, located in the Acropolis in modern Athens.
(https://commons.wikimedia.org/wiki/File:Lascar_The_Acropolis_from_Mount_Lycabettus_(4517726664).jpg.)

theaters, temples, arches, aqueducts, and monuments of varying sizes. Great engineers, the Romans were master builders.

Under the Romans, cities grew in much of southern and western Europe in a process that diffused from the southeast to the northwest, leaving a huge impact on the subsequent human geography of Europe. Examples of Roman cities, some established on Celtic sites, include Lyon and Paris in France; Londinium (London) in Britain; Belgrade in Serbia; Vienna, Austria; Trieste, Italy; and Toledo, Spain (Fig. 2.24). But by far the largest of these was Rome itself, which at its height may have had upwards of a million people. Enormous aqueducts channeled large volumes of water to the city from surrounding hills. In Rome, some of the largest and most famous monuments and temples were to be found, such as the Forum, the Colosseum (Fig. 2.25), and the Pantheon.

Eventually, the Roman Empire collapsed. Many factors contributed to its fall. Famed 18th century historian Edward Gibbon blamed the growth of Christianity, which became the state

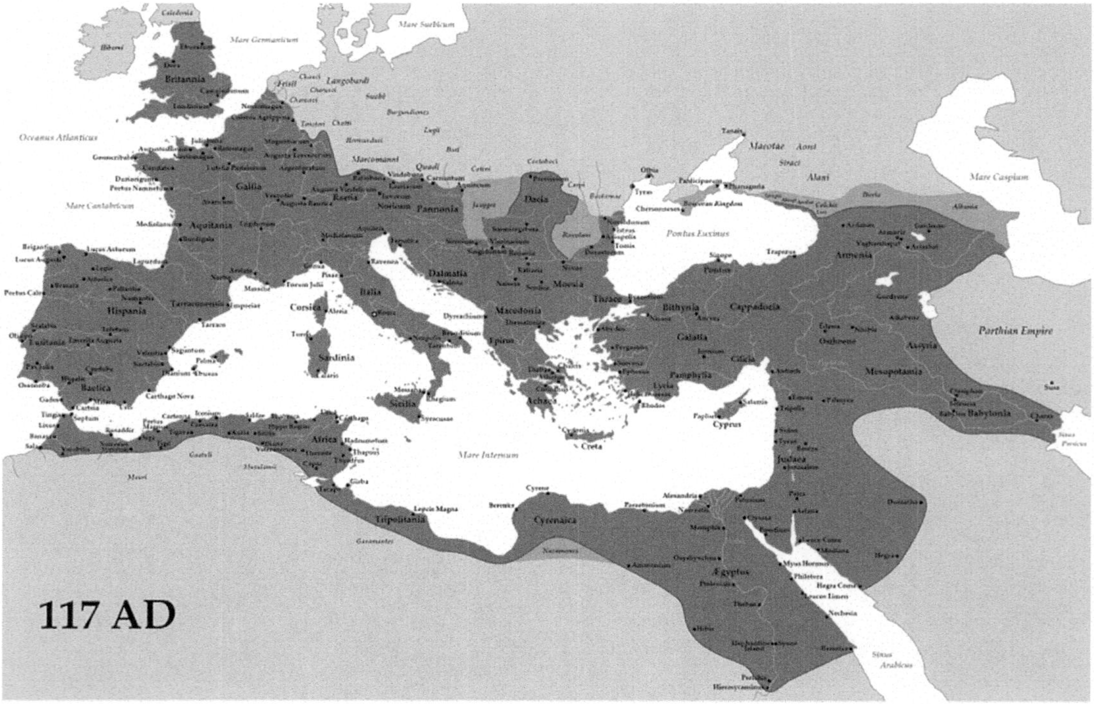

Figure 2.23
The Roman Empire at its height, in the 1st century AD, stretched from Scotland to Iraq, and was the largest slave-based social formation in history.
Map lines delineate study areas and do not necessarily depict accepted national boundaries.
(https://commons.wikimedia.org/wiki/File:Roman_Empire_Trajan_117AD.svg.)

religion in the third and fourth centuries. Shortages of precious metals led to gradual inflation and long-term debt. Overspending on the military sapped investments in other spheres. Several civil wars wracked the empire and destabilized it. Slave rebellions, such as that of Spartacus, presented a constant threat. And finally, attacks by Germanic tribes to the north, and also by the Asian Huns, succeeded in dismantling the empire in the late fifth century. After Rome fell, average city size in Europe declined for the next millennium.

2.3 Chapter Summary

Cities are relatively new to human beings: more than 95% of our existence was spent as nomadic hunters and gatherers during the long millennia of the Paleolithic. Roughly 10,000 years ago, agriculture was invented in several parts of the world almost simultaneously during the Neolithic Revolution. Agriculture is one of the greatest creations of human beings, allowing them to feed themselves more easily and supporting denser populations than hunting and gathering, although at the expense of endless toil in the fields. With agriculture, the

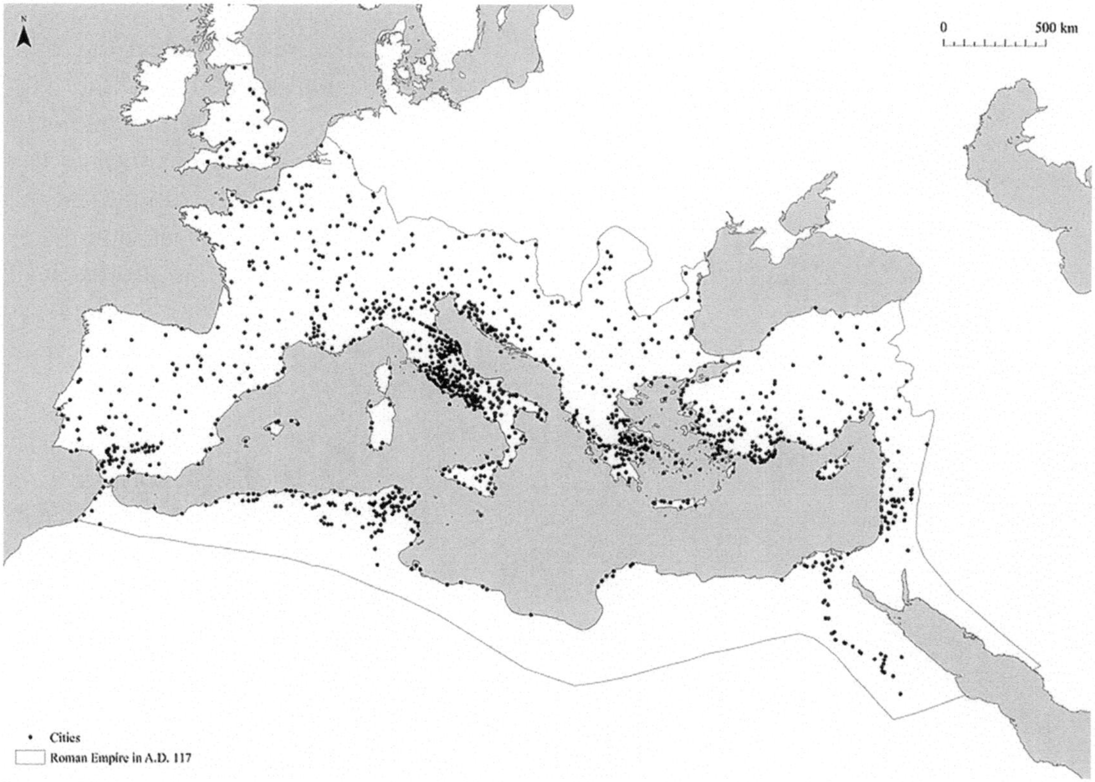

Figure 2.24
The distribution of cities founded by Romans.
Map lines delineate study areas and do not necessarily depict accepted national boundaries.
(https://commons.wikimedia.org/wiki/File:The_cities_of_the_Roman_world_in_the_Imperial_period.jpg.)

endless years of nomadism gradually came to an end, replaced by small, settled communities, villages, and eventually towns. True cities emerged much later. The emergence of agricultural societies also saw the creation of a social surplus that allowed the division of labor to become more specialized, giving rise to class-based societies. The first type of class society was slavery, which lasted for thousands of years. On the heels of agriculture human beings also created the state, organized religion, weaving, pottery, and metalworking.

The world's first cities arose in several areas in which agriculture had been present. Rivers played a key role in many places. Mesopotamia, now Iraq, is widely regarded as the first place where this process began, on the banks of the Tigris and Euphrates Rivers. Egypt was a close second, with cities stretched up and down the Nile River. In South Asia, the Indus River played a similar role in what is now Pakistan. In northern China, Neolithic cultures gave rise to cities along the banks of the Huang He River. Cities also arose in the Americas, including Mesoamerica (Mayans, Aztecs), South America (Incas), and even in a few places in North

Figure 2.25
The Coliseum, in Rome, was the largest building ever erected by the Romans.
(https://commons.wikimedia.org/wiki/File:Colosseo_2020.jpg.)

America (Cahokia). In Africa, sophisticated urban cultures arose in West Africa along the Niger River, Ethiopia, and Zimbabwe. Early urbanization in Europe was greatly influenced by Middle Eastern cultures, and included the Bronze Age civilization of the Minoans on Crete, and, later, the city-states of ancient Greece. The Roman Empire, the largest civilization before the Industrial Revolution, marked the apex of these slave-based social formations, and initiated cities in much of southern and western Europe.

Study Questions

1. What are five consequences associated with the Neolithic Revolution?
2. What is the relationship between a surplus and the rise of class-based divisions of labor?
3. What role did river valleys play in the rise of cities?
4. What did an agricultural surplus have to do with the rise of cities?
5. Where is the Fertile Crescent?
6. On a world map, show five places where cities originated.

Further Reading

Guillermo Algaze, Burchard Brenties, A.Bernard Knapp, et al., The Uruk Expansion: Cross-cultural Exchange in Early Mesopotamian Civilization [with Comments and Reply]. Current Anthropology. 0011–3204. 30 (5) (1989), 571–608, Available from: https://doi.org/10.1086/203786.

Maarten Bosker, City origins. Regional Science and Urban Economics. 01660462. 94 (2022), 103677, Available from: https://doi.org/10.1016/j.regsciurbeco.2021.103677.

Diane Z. Chase, Arlen F. Chase, William A. Haviland, The Classic Maya City: Reconsidering the "Mesoamerican Urban Tradition.". American Anthropologist. 0002–7294. 92 (2) (1990), 499–506, Available from: https://doi.org/10.1525/aa.1990.92.2.02a00210.

Mark W. Chavalas, Marc van de Mieroop, The Ancient Mesopotamian City. Journal of the American Oriental Society. 00030279. 119 (3) (1999), 520, Available from: https://doi.org/10.2307/605965.

V.Gordon Childe, The Urban Revolution. Town Planning Review. 0041–0020. 21 (1) (1950), 3, Available from: https://doi.org/10.3828/tpr.21.1.k853061t614q42qh.

C. Coquery-Vidrovitch, The Process of Urbanization in Africa (From the Origins to the Beginning of Independence). African Studies Review. 15552462. 34 (1) (1991), 1–98, Available from: https://doi.org/10.2307/524256.

George L. Cowgill, Origins and Development of Urbanism: Archaeological Perspectives. Annual Review of Anthropology. 0084–6570. 33 (1) (2004), 525–549, Available from: https://doi.org/10.1146/annurev.anthro.32.061002.093248.

Sean Fox, Urbanization as a Global Historical Process: Theory and Evidence from sub-Saharan Africa. Population and Development Review. 0098–7921. 38 (2) (2012), 285–310, Available from: https://doi.org/10.1111/j.1728–4457.2012.00493.x.

P. Hall, Cities in Civilization. (2001).

Mario Liverani, Zainab Bahrani, Marc Van De Mieroop, Uruk: The First City. Equinox Publishing Ltd (2006). Available from: https://doi.org/10.1558/isbn.9781845534875.

C. Maisels, Early Civilizations of the Old World. (1999).

M. Mann, The Sources of Social Power. (1986).

N. Morley, Metropolis and Hinterland: The City of Rome and the Italian Economy, 200 BC-AD 200. (1996).

Pollock, Ancient Mesopotamia. (1999).

E. Soja, Putting Cities First: Remapping the Origins of Urbanism. A Companion to the City. (2000).

Peter J. Taylor, Extraordinary Cities: Early 'City-ness' and the Origins of Agriculture and States. International Journal of Urban Regional Research. 0309–1317. 36 (3) (2012), 415–447, Available from: https://doi.org/10.1111/j.1468–2427.2011.01101.x.

A. Thomas, The Evolution of the Ancient City: Urban Theory and the Archaeology of the Fertile Crescent. (2010).

P. Wheatley, The Origins and Character of the Ancient Chinese City: Volume 2, The Chinese City in Comparative Perspective. (2017).

CHAPTER 3

Historical Development of Western Cities

Chapter Outline

The urban geography of the West—Europe and North America—reflects the accumulated impacts of a long series of historical processes and changes, including economic, political, and cultural transformations. This chapter sketches the development of Western cities in several stages. It opens by examining medieval cities during the long era of feudalism. Next it turns to the Industrial Revolution and its multitudinous impacts on the city. The third part focuses on the historical development of American cities from precolonial times to the present.

3.1 Medieval Cities

Prior to capitalism, the prevailing form of economic and social relations in Europe was feudalism, which lasted from roughly the sixth to the sixteenth centuries (persisting later in some places, even into the early twentieth century in parts of Eastern Europe). After the Roman Empire's collapse in the fifth century, many aspects of feudalism emerged from the Frankish Empire that dominated present-day France, Germany, and Italy from the seventh to the tenth centuries. Feudal society remained relatively stable for a long period sometimes called the Middle Ages (situated between classical slavery and capitalism). Feudalism was not unique to Europe, but existed in similar form in Japan and (to some extent) India.

Urban Geography. DOI: https://doi.org/10.1016/B978-0-443-30281-7.00003-7

Feudalism's political, cultural, and economic characteristics made it qualitatively different from capitalism. Compared to the dynamic, ever-changing world of market-based societies, feudalism produced a remarkably stable and conservative world that changed slowly. Most people's lives—working the land in a cycle of endless drudgery—were the same from one generation to another.

The ruling class of feudalism was an aristocratic nobility whose power lay in the ownership of land. Many tiers existed within this ruling class, including lords, dukes, earls, barons, and others. In an overwhelmingly rural society in which agriculture produced a relatively small amount of food per unit of land, peasants and farmers formed the vast majority of the population. Aristocrats typically owned large estates of farmland. In the manorial system, the extraction of wealth from peasants and farmers occurred through the payment of rent, in which tenant farmers paid a large share, perhaps half, of their output to the local lord. The aristocracy controlled the government, including the military and penal systems. Knights and the military existed to enforce the rule of aristocratic law and to protect communities from brigands, robbers, and invaders.

Farming under feudalism was based on animate sources of energy, that is, human and animal muscle power. Peasants and draft animals worked the fields, collected firewood, drew water from wells, and performed innumerable other tasks. Because child labor was widely used on the farms, birth rates were high (meaning women were pregnant most of the time). The population lived in small hamlets and villages, producing their own food, clothing, and other necessities. Religion was the dominant force in culture. Peasants were illiterate, ignorant of events even a few miles away, and unaware even of what century they lived in.

Although feudal society was predominantly rural and agricultural, a few cities and towns existed. Urban areas under feudalism differed from those of today in many ways. Because agricultural productivity rates were low, and thus so was farmers' abilities to support city dwellers, cities were small. Most hamlets did not exceed 200 or 300 people, and cities over 10,000 people were rare (Fig. 3.1).

Feudal cities were densely populated, with the inhabitants crowded together, often in very unsanitary conditions. The streets were often covered with mud and animal waste. The centers of feudal cities frequently consisted of a walled fortress (Fig. 3.2), often with a palace located within where the local lord lived, surrounded by concentric rings of protective walls. Because land was not a commodity to be bought and sold, but allocated on the basis of power and status, there was little differentiation among land uses: without land markets to separate land uses according to their profitability, commercial and residential land uses were mixed together.

Within the cities, feudal **guilds**, or associations of craft workers and artisans, produced a variety of goods. The primary functions of guilds were to limit competition by keeping

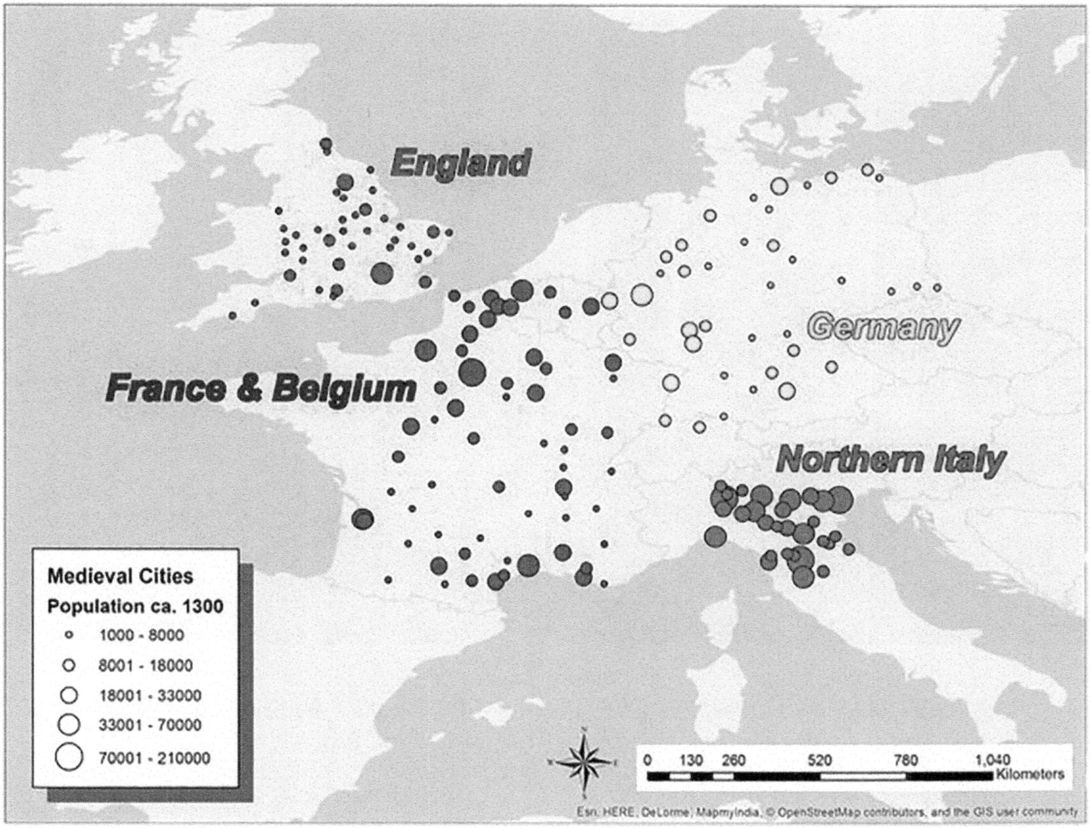

Figure 3.1
Map of European Cities in 1300. Map lines delineate study areas and do not necessarily depict accepted national boundaries.
(https://www.researchgate.net/publication/308910090_Population-Area_Relationship_for_Medieval_European_Cities/figures?lo=1.)

imports out as well as to train new generations of labor. Guilds consisted of skilled workers with years of experience, and were organized by the type of good they produced. There were, for example, blacksmiths' guilds, weavers' guilds, goldsmiths' guilds, and guilds for bakers, leather workers, papermakers, glass workers, and shoemakers. Young men who were chosen to work in the guild spent years as apprentices learning the trade before becoming craftsmen in their own right.

3.2 The Industrial Revolution and the City

The pace of economic, technological, social, and geographic change accelerated greatly in the 18th and 19th centuries during the Industrial Revolution. Note that the Industrial Revolution occurred long after capitalism began; indeed, for most of capitalism's history, it involved

Figure 3.2
Carcassonne, France, is a good example of the walled medieval city.
(https://commons.wikimedia.org/wiki/File:1_carcassonne_aerial_2016.jpg.)

preindustrial labor-intensive artisanal and household-based forms, of production. However, starting in the mid-1800s, an explosive increase in the speed and productivity of capitalist production in Europe, North America, and Japan transformed the worlds of work, everyday life, and the global economy. In the 20th century, industrialization spread to Eastern Europe and the Soviet Union, and since the 1970s, it has spread to selected parts of the developing world, particularly East Asia.

Three elements of industrialization are particularly important: the use of inanimate energy, technological changes, and productivity growth.

3.2.1 Inanimate Energy

If preindustrial societies relied on animate sources of energy (i.e., human and animal muscle power) to get things done, industrialization can be defined as the harnessing of inanimate sources of energy. This process marked a major milestone in human history. Several types of inanimate energy were tapped over time. The first involved running water moving from higher to lower elevations. This source had been used even in the late medieval era in mills to grind grain into flour. Running water was a major source of energy in the earliest stages of the Industrial Revolution, but it constrained firms to locations near streams and rivers. Many early textile plants in Britain and New England, for example, used this strategy. There were severe drawbacks to this approach: Many streams dry up in the summers, and locating one may put the producer inconveniently far away from the market.

A more efficient source of inanimate energy involves the steam engine, for which the designs were created by James Newcomen in 1712; the first operating model was built by the Scottish engineer James Watt in 1769. This technology marked a turning point in the process of industrialization. Originally designed to pump seawater out of underwater coal mine shafts that penetrated coastal coal layers that extended under the ocean, the steam engine could do the work of dozens of men far more efficiently. Wood provided the first major source of fuel for this invention, which required heating water into steam to drive the engine's pistons. As the demand for wood led to forests in Britain being cut down in large numbers, wood supplies dwindled, and the rising cost eroded profits. (Note the steam engine itself did not generate the energy; that came from fuels such as wood or coal.)

As wood became scarce, producers switched to coal, which was mined in large quantities. Thus, as Britain industrialized, several areas became major coal-producing centers, including Wales and Newcastle. As the Industrial Revolution spread across Europe, the large coal deposits of the northern European lowlands became increasingly important, including in northern France, Belgium, the Ruhr region in Germany, and Silesia in southern Poland. This relation exemplifies how nature helps to shape the formation of geographies. In the United States, coal deposits in Appalachia played a key role in the nation's industrialization.

In the 19th century, coal was joined by other fossil fuels, particularly petroleum and, to a lesser extent, natural gas. The abundance of cheap energy was the lifeblood of industrialization, and production processes became increasingly energy intensive as a result. This substitution of inanimate for animate energy freed tens of millions of people from drudgery and allowed large numbers to live relatively comfortable lives in the expanding middle class that industrialization produced.

3.2.2 Technological Innovation

Capitalism involves constant change. Firms, under the lure of profits and threat of ruin, engage in innovation as a way to reduce costs and increase revenues. Although innovations certainly emerged prior to industrialization, the Industrial Revolution witnessed a dramatic jump in the number, diversity, and applications of new technologies. A technology is a means of converting inputs to outputs. These can range from extremely simple (e.g., a digging stick) to highly sophisticated (e.g., a computer). As industrialization produced an increasingly complex division of labor based on mass production, new inventions sprouted up rapidly.

During the Industrial Revolution, a major reorganization in the nature of labor occurred with the development of the factory system. Prior to this era, production was organized on a small-scale basis, including home-based work. By the late 18th century, firms in different industries began grouping large numbers of workers together under one roof, a process that effectively created the industrial working class. Never before in human history had so many workers been concentrated on a permanent basis, a feature that changed how they lived and viewed each other, and themselves. Inside factories, workers used many types of machines representing vast amounts of capital. The introduction of interchangeable parts, invented by American gun-maker Eli Whitney, made machines more reliable and easier to fix (Whitney also invented the cotton gin). By the early 20th century, Henry Ford introduced the moving conveyor belt, which further accelerated the tempo of work and the ability of workers to produce.

3.2.3 Productivity Increases

Because of the massive technological changes of the Industrial Revolution, productivity levels surged. Productivity refers to the level of output generated by a given volume of inputs. For example, a farmer may measure productivity in terms of bushels per acre (or metric tons per hectare), or a firm might measure it in terms of units produced per unit of labor hour. Productivity increases refer to rising levels of efficiency, such as greater levels of output per unit of input (e.g., labor hour or unit of land) or, conversely, fewer inputs per unit of output.

Productivity levels rose exponentially in the 19th century, generating several important effects. As the cost of producing goods declined, standards of living rose. Most workers labored long hours under horrific conditions and endured standards of living still quite low compared with those we enjoy today. But nonetheless, over several decades, industrialization saw many kinds of goods become increasingly affordable, meaning that the working class gradually became better off. Clothing, for example, which was scarce before the Industrial Revolution, became relatively cheap. As agriculture became mechanized, food supplies got cheaper and more varied, malnutrition declined, and famines in Europe essentially disappeared (except for the Irish Potato Famine of the 1840s).

3.2.4 The Urban Geography of the Industrial Revolution

Geographically, the Industrial Revolution was closely associated with the growth of cities. Almost everywhere, industrialization and urbanization have been virtually simultaneous processes. Manufacturing firms concentrated in cities, especially large ones, such as the British Midlands (e.g., Leeds, Liverpool, Birmingham, and Manchester), where the Industrial Revolution began. In North America, the US Manufacturing Belt (e.g., Pittsburgh, Buffalo, Cleveland, Detroit, Chicago, and Milwaukee) played a similar role.

The reasons why firms concentrated in cities are important. Often, there is a tendency to attribute this phenomenon to the presence of workers in urban areas. Which came first, firms or workers? Cities were clearly centers of capital investment as much as they were centers of labor. Yet, cities were very small when the Industrial Revolution began, and through agricultural mechanization (which reduced rural job opportunities) and rural-to-urban migration, the urban labor supply was produced.

Firms have powerful reasons to concentrate, or agglomerate, in cities. Most firms benefit substantially by having close proximity to other firms, including suppliers of parts and ancillary services. Concentration allows firms to reduce transportation costs, share infrastructure, obtain specialized information, and generate and develop a labor force with necessary skills for the field. By locating near one another, firms become more productive than if they are alone. The cities of the 19th century were composed of dense webs of industrial firms, with intricate input and output relations tying them together. Companies in these cities bought parts, services, and information from, and sold their products to, other

MAP 19.2 The Industrial Revolution

Western Civilizations, Sixteenth Edition
Copyright © 2008 W. W. Norton & Company, Inc.

Figure 3.3
Urbanization and industrialization in Europe during the Industrial Revolution. Map lines delineate study areas and do not necessarily depict accepted national boundaries.
(From https://ourworldindata.org/grapher/long-term-urban-population-region?tab=map.)

companies located nearby. In short, industrialization led to the creation of large urbanized manufacturing complexes across Europe (Fig. 3.3).

Industrialization changed everyday life in many ways. The switch to inanimate energy made many products, including food and clothing, much cheaper. Despite widespread poverty, standards of living began to rise rapidly, creating a working class and then a middle class with disposable incomes. As the need for children to work on the farms declined, birth rates fell and families became smaller. The factory system brought large numbers of workers together for the first time, and the introduction of machinery greatly changed the nature of work.

Industrialization changed societies from predominantly rural to increasingly urban in character. In Europe, North America, and Japan, for the first time in history the majority of people lived in cities. The growth of cities in industrial societies is frequently depicted using an urbanization curve, which illustrates the percentage of people living in urban areas over time. In the United States, for example, the first national census of 1790 showed that 95% of Americans lived in rural areas. This proportion decreased throughout the 19th century, and, by 1920, 50% of the nation's population lived in cities. Today, it is roughly 85%. This pattern is similar in every other country that has industrialized.

Cities during the Industrial Revolution were generally not pleasant places to live (Fig. 3.4), especially for the vast numbers of the impoverished working class. Smoke and pollution from factories and fires often rendered the air unbreathable. Water supplies were frequently

Figure 3.4
Sketch of a British city during the Industrial Revolution.
(https://www.historyskills.com/classroom/year-9/urbanisation-reading/.)

contaminated, leading to the spread of diseases such as cholera. The streets were usually filthy, covered with trash and horse droppings, and rodent populations were widespread. For these reasons, life expectancy in such environments was low, often under 40 years.

3.3 The Historical Geography of US Cities

One chapter in the long historical trajectory of Western cities centers on North America, particularly the United States. With roots in the settlements of Native Americans, cities in the United States and Canada have steadily grown in number, size, and complexity.

3.3.1 Preindustrial American Cities

The first cities in North America were Indigenous ones (Chapter 2), including Cahokia, the settlements that lined the Mississippi and Ohio Rivers, and the peoples who inhabited the deserts of the Southwest. All of these disappeared, for various reasons, before the arrival of the Europeans.

In the 16th century, when Spain was the world's most powerful empire, the Spanish colonized Florida. It was here that the city of St. Augustine was founded in 1565. There is no doubt that the city was an important colonial port, and vestiges of the Spanish era are imprinted on its landscape (Fig. 3.5). But St. Augustine's claim to be the "nation's oldest city" is belied by the facts. First, Indigenous cities were much older. Second, Spanish colonial cities such as San Juan, Puerto Rico (which is part of the United States) were established earlier.

The Spanish influence was not confined to Florida, as Spanish conquistadors conquered much of the Southwestern United States as well. Throughout the 17th, 18th, and 19th centuries, the Spanish founded numerous settlements in California called presidios. These stretched up the Pacific coast from San Diego to San Francisco (Fig. 3.6), and served as centers of administration, military supply points, and, equipped with churches, missionaries, and priests, places to disseminate the Catholic faith. The largest of these was Nuestra Ciudad de la Reina de Los Angeles, or Our City of the Queen of Angeles, founded in 1781.

In the early 17th century, European powers initiated several settlements on the East Coast (Fig. 3.7). The Dutch founded New Amsterdam in what is now New York. The French colonized Quebec and the St. Lawrence River and sent fur traders down the Mississippi River. But it was the British who established the largest colonies, including plantations in Virginia and the Carolinas. The Jamestown, Virginia settlement dates back to 1607, and was followed by several others. In 1620 the first ship arrived at Plymouth Rock, in Massachusetts, beginning a series of settlements motivated by a quest for religious freedom as they were populated by Pilgrims, Quakers, and Congregationalists.

Figure 3.5
Spanish colonial fort in St. Augustine.
(https://commons.wikimedia.org/wiki/File:Old_City_Gate,_St._George_Street,_Colonial_Quarter,_St._Augustine,_FL.jpg.)

Many cities were founded at the mouths of rivers, including New Amsterdam (later New York) on the Hudson River, Boston on the Charles River, and Philadelphia on the Schuylkill River. These locations testify to the importance of maritime trade in the formation of these urban areas. These cities served as beachheads for the expansion of capitalism, with trading links to their home countries. Most were centers of petty commodity production, in which artisans using animate (living) energy made a variety of products for local consumption, including farm tools, clothing, and household goods. These cities also exported timber, furs, grains, fish, and other goods, and imported manufactured goods, mostly to Britain. Most became centers of the slave trade, which began almost immediately with European colonialism. The first slaves were brought from Africa in 1619. In connecting Europe, Africa, and North America, they became part of the North Atlantic trading network known as the Triangular Trade system, in which Britain sold guns, whiskey, and cloth to African kings,

Figure 3.6

Spanish presidios in California. Map lines delineate study areas and do not necessarily depict accepted national boundaries.
(https://en.wikipedia.org/wiki/List_of_Spanish_missions_in_California#/media/File:SpanishMissionsinCA.png.)

who sold their own people as slaves, and the plantations in the Caribbean and the American South that used slaves to export cotton, tobacco, rum, and molasses to Britain.

American urbanization was shaped by several forces that combined to produce a unique urban system. As the country grew in size and power over time, the United States became enormously wealthy; its strong competitive position internationally, including trade and

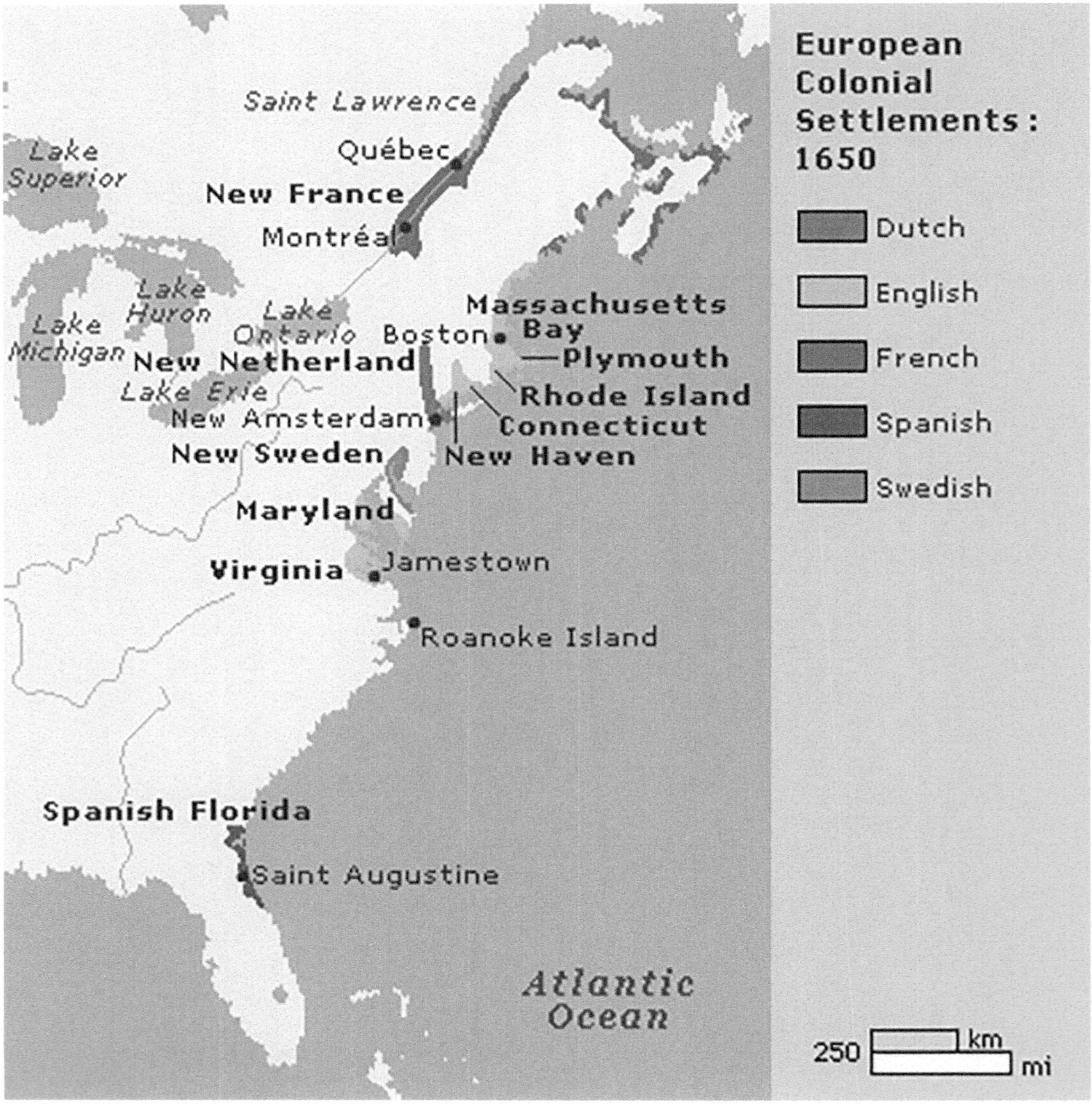

Figure 3.7
European settlements in Eastern North America, 17th century. Map lines delineate study areas and do not necessarily depict accepted national boundaries.
(From https://commons.wikimedia.org/wiki/File:NorthAmerica1762-83.png.)

international alliances, helped to fuel processes of urban growth at home. By the close of the 19th century, the United States had become the world's largest national economy, with ample job opportunities that attracted wave upon wave of immigrants. It enjoyed high degrees of social mobility that drove the suburban expansion that has long been a part of the country's urban fabric. Starting in the early 19th century, as European-Americans began to move westward in large numbers, the mythologized "frontier" became the edge of capitalist development as it unfolded over the continent. Transportation systems such as the railroad and steamship played a major role in this urban expansion. Finally, despite popular misunderstandings of the so-called "free market," in reality the state, or government, helped to fuel urban expansion, subsidizing railroads, providing public services, and, in the mid-20th century, building networks such as the Interstate Highway System.

New York was a major center of early American urban development, fiercely competing with Boston and Philadelphia. Originally, it was a Dutch colony called New Amsterdam. The Dutch purchased the land from the Lenape and Manhattan tribes, to whom the idea of buying and selling land was foreign. Its first governor was Peter Stuyvesant, a one-legged veteran who oversaw the new colony. The Dutch built Wall Street to protect them from the Indigenous residents, and remnants of the Dutch legacy are found in place names like Brooklyn and Harlem (originally "Breukelen" and "Haarlem"). In 1664 the British took over the Dutch colony and renamed it New York, in honor of the Duke of York.

Cities were the exception, not the norm, in early America. The country was overwhelmingly rural, and the bulk of the labor force worked on farms, testimony to the use of animate energy and preindustrial agriculture. In the first US census of 1790, only five percent of people lived in urban areas, mostly along the eastern coast.

The largest cities of the newly formed United States were Boston, Philadelphia, and New York, although other important centers included Newport, Rhode Island; Baltimore, Maryland; and Charleston, South Carolina. Boston was a major center of shipping, trade, and banking, notably maritime insurance. Although its colonial population was less than 30,000 people, it swelled during the 19th century, fueled by immigration and the Industrial Revolution.

Among the new republic's premier cities, the two largest, New York and Philadelphia, were engaged in a fierce competition to open up new lines to the lucrative domains west of the Appalachian Mountains. This region, stretching along the southern parts of the Great Lakes to Kentucky and Tennessee, offered immense amounts of farmland, timber, and mineral ores. New York won this contest decisively through the construction of the Erie Canal (1817–1825), which cut across New York State and linked to the Hudson River (Fig. 3.8). The canal not only gave New York access to midwestern wealth, firmly establishing it as the country's premier city, it also gave rise to a series of smaller urban areas that prospered with the growth of traffic on the canal, including Albany, Syracuse, Utica, Rochester, and Buffalo. Many immigrants worked on the canal, and it facilitated immigration to the Midwest and West.

In 1800, the United States established Washington, DC, as its national capital (earlier capitals included New York and Philadelphia). The District of Columbia was created out of land taken

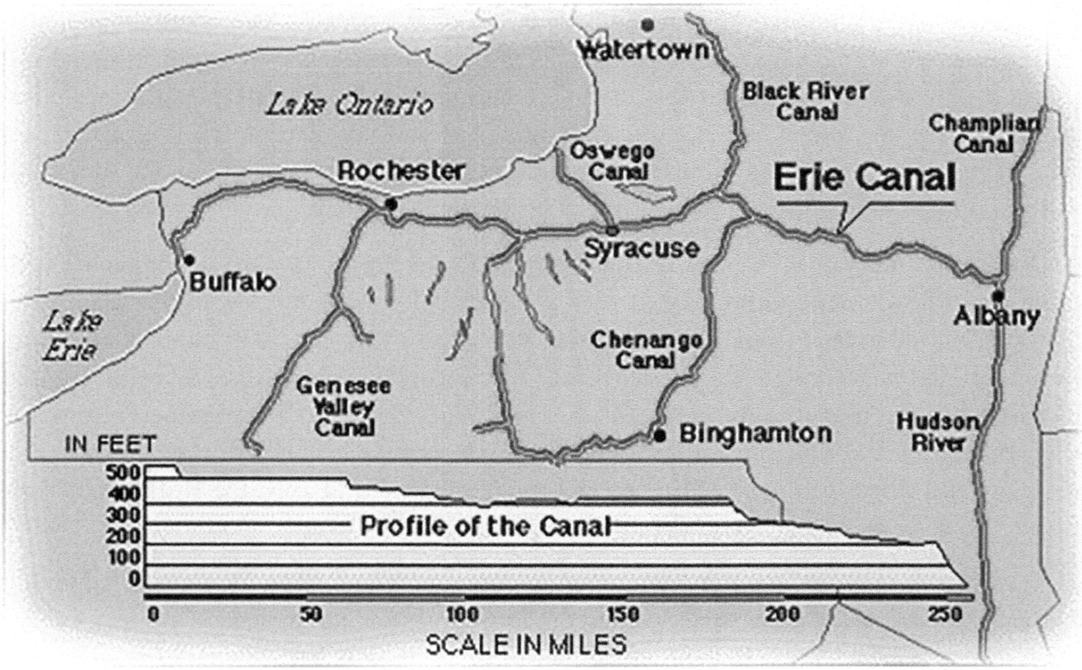

Figure 3.8
Map of the Erie Canal. Map lines delineate study areas and do not necessarily depict accepted national boundaries.
(From New York State, https://commons.wikimedia.org/wiki/File:Erie-canal_1840_map.jpg.)

from Virginia and Maryland. The new city was carefully designed and planned by Pierre Charles L'Enfant, a French-American artist and engineer. He created the L'Enfant plan for George Washington, a landmark of urban design, which laid out broad avenues, plazas, public spaces, the president's house (now the White House), and the National Mall (Fig. 3.9). During the War of 1812, the British burned most of the original buildings, which subsequently had to be rebuilt.

Cities in the early American South tended to be smaller and less numerous than their northern counterparts, largely because of the region's dependence on agriculture. Charleston was long the most important Southern urban area, in large part because of the slave trade. Other early Southern cities included Wilmington, North Carolina, and Savannah, Georgia. As the region grew in the 19th century, a series of small cities nestled against the foothills of the Appalachian Mountains became known as "fall line" cities, including Richmond, Virginia; Raleigh, North Carolina; Columbia, Augusta; and Macon, Georgia.

In 1718, the French, who controlled the Mississippi River basin until 1804, founded the city of New Orleans, named after the French city of Orleans, at the mouth of the river in Louisiana. After the Louisiana Purchase in 1804, the city became part of the United States.

Figure 3.9
L'Enfant's plan for Washington, DC.
(https://commons.wikimedia.org/wiki/File:L%27Enfant_plan.svg.)

The city became an enormously important port, funneling grains, meat and other commodities from the Mississippi River region to the East Coast and to Europe, and remains one of the largest ports in the world today. New Orleans has long served as the center of French-American culture, as exemplified in its French Quarter and the Cajun dialect, music, and cuisine. Today its economy revolves largely around tourism.

American cities before the mid-20th century were often unpleasant places to live. Access to clean water was often difficult; many people resorted to lightly alcoholic fruit juices, which were at least sanitary. Water is a major vector for infectious diseases, so outbreaks of cholera were not uncommon. Cities with numerous horses had streets plastered with horse manure. Waste removal was intermittent, with trash accumulating on the streets, providing fertile ground for rodents and pests.

Because the vast majority of buildings were built of wood, which was cheap, fire posed an ever-present danger. Fires were common in European cities as well, such as the Great Fire in

London in 1666. Fires started for various reasons—accidents, arson, lightning strikes—and spread rapidly. Before the establishment of professional fire departments, they were fought by volunteer fire brigades. Large fires burned, for example, in New York (1776 and 1845), Boston (1787 and 1872), New Orleans (1788 and 1794), Detroit (1805), St. Louis (1849), and San Francisco (1851 and 1863). One of the most calamitous was the Great Chicago Fire of 1871 (Fig. 3.10), which killed 300 people and destroyed 17,000 structures, leaving many homeless.

3.3.2 Industrial American Cities

As the Industrial Revolution began to sweep over North America in the early 19th century, it had enormous impacts on the nature and fabric of urbanization. Industrialization changed everything, from how crops were grown to how people moved, their jobs, and the social and economic structure of urban areas.

Figure 3.10
Rendition of the Great Chicago Fire of 1871.
(https://commons.wikimedia.org/wiki/File:Great_Chicago_Fire_cph.3b20287.jpg.)

Industrialization was fueled in large part by a steady stream of immigrants, who provided cheap and willing labor. In addition to slaves—unwilling immigrants kidnapped from Africa—the workers who became the American labor force first originated from northwestern Europe, notably Britain and Germany. These largely Protestant populations became the elites of many East Coast cities. In the 1840s, the Irish, fleeing the Great Potato Famine, arrived in large numbers, the first major group of Catholics. Many faced intense discrimination. By the late 19th century, the sources of immigration shifted to southern and Eastern Europe, with large numbers of Italians, Poles, Romanians, Russians, and others coming in. Many Chinese came in the mid-19th century, working in agriculture and on the railroads, and were later joined by Japanese and Korean immigrants. Later, after World War II, the dominant sources of immigration were East Asia and Latin America.

In the early 19th century, the textile industry flourished in southern New England, the first wave of American industrialization. Cities in Rhode Island, Connecticut, and Massachusetts flourished accordingly, as did garment production in New York City. Many of the workers in the textile industry were Irish, immigrants fleeing their country's potato famine. The Irish suffered from severe anti-Catholicism and discrimination from the country's Protestant elites. In New York, Jewish workers formed the mainstay of the garment industry, notably Jews from Eastern Europe.

Arguably the most important point of entry for European immigrants was Ellis Island, a federally owned island split between New York and New Jersey. Between 1892 and 1954, more than 12 million immigrants entered through this gateway (Fig. 3.11). Some were forced to return, others quarantined for diseases, and yet others processed and allowed to enter.

As immigrants settled in cities, they formed tightly bound ethnic communities. The reasons for this are both obvious and complex. Many immigrants had very limited financial resources, compelling them to live in cheap, densely populated areas near the downtown. The famous photos of Jacob Riis, a Danish-American journalist and photographer, captured the squalor and deplorable conditions in which many immigrants found themselves (Fig. 3.12). There were also powerful cultural reasons to cluster, including access to their native language. Ethnic communities often formed self-help organizations to assist newcomers, lending them money and providing information about jobs and housing. Moreover, many immigrants faced intense discrimination and prejudice, against which ethnic enclaves offered some protection. As a result, many cities developed distinct neighborhoods such as German town, Little Italy, Little Poland, Little Russia, Chinatown, Little Tokyo, and so forth. The children of immigrants, however, grew up as citizens of the United States, and were usually fluent in English. The great stereotype about immigrants is that they joined the "melting pot" through assimilation; while there is some truth in this observation, it tends to be simplistic and exaggerated. Different groups assimilated at different paces, and some not at all. As the second generation, enjoying English fluency and education often denied to their parents,

Figure 3.11
Italian immigrants in Ellis Island, early 20th century.
(https://commons.wikimedia.org/wiki/File:Italian_immigrants_at_Ellis_Island.jpg.)

entered better-paying occupations, they often abandoned the neighborhoods of their parents for more desirable ones farther from the city center, where land costs are lower. New waves of immigrants took their place, however. This trend indicates that while ethnic enclaves are a permanent part of the American urban landscape, they are temporary as far as any individual group is concerned.

Starting in the 1850s, Germans came to the United States in large numbers. Indeed, today more Americans trace their ethnic ancestry to Germany than any other place. In the mid-19th century, Germany was engulfed in turbulent social and political changes such as the Revolutions of 1848, giving many reasons to leave. Simultaneously, steamships laden with agricultural goods from the United States that arrived in Europe often faced little demand for the return voyage, inducing them to offer low-cost trips for people. Many Germans accordingly took steamships up the Mississippi and Ohio Rivers, leading them to form "Germantowns" in cities such as St. Louis, Cincinnati, Milwaukee, Chicago, Minneapolis, Pittsburgh, and Columbus, Ohio. This phenomenon exemplifies how the dynamics of the

Figure 3.12
Immigrant life in New York as captured by Jacob Riis.
(From U.S. National Archives and Records Administration, https://commons.wikimedia.org/wiki/File:Immigrant_Children,_Ellis_Island,_New_York_-_NARA_-_6341034.jpg.)

world system, technological changes, and geopolitics shaped the economic and residential landscapes of American cities.

After the Civil War ended in 1865, the expansion of the US urban system across the Midwest began in earnest. Steamships fueled the growth of many towns along rivers, and the growing railroad network made it easier for people to move west and commodities to move east. The agricultural hinterlands of the Great Plains flourished, and cities grew as well. Telegraphy, which started in 1844, helped to reduce the uncertainty of this process as networks of telegraph wires were erected across the continent. St. Louis, Kansas City, Wichita, Oklahoma City, and Colorado Springs were part of the postwar urbanization of the Great Plains, in which towns started as frontier settlements, became meat and agricultural processing centers, and then true urban areas.

The greatest beneficiary of 19th century urban growth was Chicago. Situated on Lake Michigan, it offered easy access to water transportation such as the freighters that brought iron ore from the northern Great Lakes. The Illinois and Michigan Canal connected it to the Mississippi River. It became the hub of a dense network of rail lines. Its booming steel, meat-packing, agricultural machinery, and other industries attracted countless numbers of immigrants; in 1900, 77% of the population consisted of either immigrants or their children. By 1870, with 300,000 people, Chicago became the country's second largest city, after New York, a status that it enjoyed until the 1970s, when it fell to third place, behind Los Angeles.

On the West Coast, immigrant labor from China formed the major labor source for the railroads. Young men recruited as laborers in China arrived in large numbers, forming the nuclei of Chinatowns in Seattle, San Francisco, and Los Angeles. Many suffered from intense racism and Sinophobia.

In 1869, Chinese workers approaching from the west and Irish workers approaching from the east met at Promontory Point, Utah, to complete the Central Pacific line, the first transcontinental railroad. The railroads facilitated the extraction of grains, minerals, and timber from the West, and played the decisive role in welding the local economies of the country into a national one.

On the West Coast, the Gold Rush of the mid-19th century and the flourishing "China trade" with East Asia generated wealth and opportunities that attracted immigrants from the United States and abroad alike. San Francisco, one of many Spanish presidios, became the largest city in the western United States, a hub of trade, immigration, and banking and California's largest city until the rise of Los Angeles in the 20th century. The 1906 earthquake and fire in San Francisco devastated much of the city, contributing to its eclipse by its southern neighbor (Fig. 3.13). In the northwest, Seattle emerged as the region's hub, a city with deep roots in the timber and lumber industry that grew as the railroads made the export of wood to cities on the East Coast feasible. Indeed, the term "skid row," now meaning a derelict part of downtown, originated from the "skid road" used to roll logs down the city's hills to the harbor.

Near the end of the 19th century, the United States became the largest national economy in the world. With abundant farmland, natural resources, massive immigration, free trade, and rapid technological change, they enjoyed rapid almost continuous growth in output and productivity, the business cycle and periodic downturns notwithstanding. The Industrial Revolution created a large, relatively prosperous middle class, poverty rates declined steadily, and cities boomed. A stream of rural-to-urban migration helped to fuel this growth.

It was during this period that the Manufacturing Belt came into its own. Stretching from southern New England westward along the southern shores of the Great Lakes (Fig. 3.14), this region became the largest industrial conglomeration in the world. Industries such as steel, rubber, glass, machine tools, railroads, and agricultural implements drove much of its growth,

Figure 3.13
Photo of San Francisco after the 1906 earthquake and fire.
(https://commons.wikimedia.org/wiki/File:The_history_of_the_San_Francisco_disaster_and_Mount_Vesuvius_horror_(1906)_(14772579952).jpg.)

as did immigration from Southern and Eastern Europe. Within it, cities grew to unprecedented sizes.

The steel industry, used widely in railroads, shipping, and construction, became the economic mainstay of many Manufacturing Belt cities. Examples include Buffalo, New York; Cleveland, Youngstown, and Toledo, Ohio; Chicago, Illinois; and Milwaukee, Wisconsin. Akron, Ohio, became the world capital of rubber production. But of all the Manufacturing Belt cities, it was Pittsburgh that held pride of place. The world's largest steel-producing center, Pittsburgh became the very model of the industrial city. At the intersection of the Ohio, Allegheny, and Monongahela rivers (Fig. 3.15), it was ideally suited to obtain Appalachian coal, iron ore from the northern Great Lakes, and other inputs. Massive waves of immigrants provided the labor that made it the most successful industrial city in history.

Another city that became very important and influential was Detroit. It was there that famed industrialist Henry Ford started the newest addition to the suite of Manufacturing Belt

Figure 3.14
Map of the Manufacturing Belt. Map lines delineate study areas and do not necessarily depict accepted national boundaries.
(https://commons.wikimedia.org/wiki/File:Rust-belt-map.jpg.)

industries, the automobile industry. (The first US plant was started in Dearborn, near Detroit, in 1909.) Ford's model of mass producing cars revolutionized not only automobile production but many other industries as well. In adopting the moving assembly line and a detailed division of labor in which jobs were highly specialized, Ford achieved unprecedented levels of productivity growth. As a result, automobiles could be produced cheaply and made affordable to the middle class. The auto industry and parts makers became the largest sector in the Manufacturing Belt, the backbone of the Detroit economy as well as many other cities in the region. By the 1950s, one out of five jobs in the United States revolved, directly or indirectly, around auto production.

The political organization of American cities in the late 19th and early 20th centuries centered on what became known as political "machines," quasiformal organizations that at times resembled organized crime families that took control of the reins of government in many cities. The machines controlled public employment and contracts, as well as the courts and judicial systems. Corruption was rampant. The machines had strong roots in immigrant communities, many of which felt excluded by traditional labor unions. The machines provided

Figure 3.15
Image of Pittsburgh at its economic zenith, in 1904.
(https://commons.wikimedia.org/wiki/File:Thaddeus_M._Fowler_-_Pittsburgh,_Pennsylvania_1902.jpg.)

immigrants with jobs, housing, and naturalization assistance in return for votes. At the top was the machine "boss," under whom worked ward captains and others who made sure the city worked to their advantage. The largest and most famous example was Tammany Hall in New York, run by the infamous Boss Tweed, which effectively governed the city until the mid-20th century. Other examples included the Rizzo machine in Philadelphia, the Daley machine in Chicago, and the Pendergast machine in Kansas City. Although the machines began to lose power during the Progressive Era (1890s–1920s), when civil service exams and citywide elections were introduced, many exerted an enduring influence until the 1970s.

The Progressive Era reforms were part of a broader effort to make cities more livable. In addition to curtailing the machines, the period saw urban reformers introduce child labor laws and attempts to clean up the tenement buildings that housed people living in poverty. The YMCA created libraries and gyms, and the Salvation Army started soup kitchens for people who needed them. It was the period of muckraking journalism that exposed corruption. Nationally, this saw attempts to rein in the power of large corporate oligopolies, such as with the antitrust movement. Direct elections of senators replaced their appointment by governors.

By 1920, when the United States became predominantly urbanized for the first time, it also saw women given the right to vote in presidential elections.

Although this period is known as the heyday of manufacturing, during the early 20th century services became an increasingly important part of the US economy. Indeed, measured by employment, the United States became a postindustrial society even when manufacturing was the driving force of the economy. The growth of services, including many white-collar professional jobs in management, finance, engineering, marketing, public relations, and accounting, led to a proliferation of office towers in downtown areas, replacing the smokestacks that had been the ultimate sign of industrial prowess. In every large city, downtowns saw clusters of office buildings (Fig. 3.16). This process was greatly enabled by innovations such as structural steel, which allowed higher buildings; mass transit systems such as subways that shuttled workers in and out of cities on a daily basis; electric lighting; and the telephone and the elevator, which made vertical movement easier.

Figure 3.16
New York in 1922.
(https://commons.wikimedia.org/wiki/File:Old_New_York_yesterday_and_today_(1922)_(14594492839).jpg.)

By 1920, the United States had become a predominantly urban country. This simple fact carried enormous implications. Life in cities was different than in rural areas, far more dense and crowded. Many people suffered from urban poverty. Streams of rural-to-urban migrants, as well as immigrants from abroad, flooded American cities. The infrastructure struggled to keep pace. Cities began to grow in size, power, and wealth, affecting the national political agenda.

3.3.3 The History of American Urban Transportation

Central to the organization, structure, growth, and success of cities is transportation, the movement of people within and among urban spaces. Transportation is how humans conquer space. Its importance cannot be overemphasized. The modes of transport that developed historically in cities show great changes over time, although generally they led to systems that were faster, safer, and could move more people with increasing efficiency.

For the vast bulk of human history—that is, before the Industrial Revolution—transportation relied on animate sources of energy. For most people, that meant walking. Daily commutes to and from fields were the norm, and even during the early industrial era most people walked to their jobs. Consequently, urban streets and sidewalks were often crowded. This left many exhausted by the time they returned home.

For the wealthy, preindustrial transportation meant horses, although few could afford them. (Hence the Spanish word for gentlemen, *caballero*, meant one who owned a *caballo*, or horse.) In the 19th century, many people relied on the horse-drawn omnibus, a wagon pulled by a team of horses that served as a primitive form of mass transportation (Fig. 3.17), although the companies that ran these were privately owned. The omnibus was the first widely used form of mass transportation, and reflected the growing divide between home and work that characterized rapidly growing cities.

In the late 19th century, the introduction of the electric streetcar revolutionized urban transportation. It was among several transportation technologies to harness inanimate energy (others include the railroad, automobile, and airplane). Streetcars provided affordable transportation to the growing middle class and greatly expanded the size of commuting fields (Fig. 3.18). Streetcars were also a major force in the growth of early suburbs, as those who could afford single-family homes on the urban periphery began to relocate to outlying areas.

Starting in the late 19th century, several large cities began to construct subway systems. Subways are expensive to build and operate, and therefore require high volumes of riders to make them cost-effective. For this reason, subways are only found in metropolitan areas. The London Underground, which began in 1863, is the world's first and oldest mass transit system. The New York subway system began around 1901 (Fig. 3.19). Today, the New York

Figure 3.17
Horse-drawn omnibus in London, 1865.
(https://commons.wikimedia.org/wiki/File:London_omnibus_in_1865.jpg.)

subway system is the world's largest; begun in the late 19th century, today it carries more than 3.2 million passengers daily. Many subway lines were initiated by private corporations and then later were purchased and operated by municipal governments. Subways became important sources of transportation for the vast numbers of office and service sector workers who came to dominate the labor markets of large downtown regions.

The invention and subsequent mass adoption of the automobile greatly revolutionized urban landscapes in North America, Europe, and later, much of the rest of the world. The automobile was a European invention, but it was the American industrialist Henry Ford who pioneered the mass production of cars, making them affordable to the middle class. Ford started the first US automobile factory in Dearborn, Michigan, in 1909. Auto ownership in the United States began to grow quickly.

Rising automobile ownership greatly enhanced urban mobility and had enormous impacts on city landscapes. In the early 20th century, many cities began to build parkways to allow for faster movement, although many became quickly congested. After World War II, parkways gave rise to freeways, including the Federal Interstate Highway System (Fig. 3.20).

Figure 3.18
Streetcars in Los Angeles, 1908.
(https://commons.wikimedia.org/w/index.php?search=streetcars+historical&title=Special:MediaSearch&go=Go&type=image.)

Los Angeles emerged as the prototypical automobile-based city, with multiple centers sprawling over vast distances. Several automobile companies purchased the light rail systems that had been in place in an effort to force Americans to drive. Automobile ownership quickly became a sign that one had joined the middle class.

Like the streetcars before them, parkways and freeways made access to the urban periphery even easier, spurring several more rounds of suburbanization. This process led to an exodus of the middle class from central cities, a phenomenon that began with streetcars, accelerated in the 1920s, and exploded after World War II. Finally, as cars became ubiquitous, urban landscapes saw the rapid growth of parking lots, gas stations, car washes, and automobile repair shops.

Not everyone benefited from this transformation. Light rail companies began to go bankrupt in the face of competition with cars and trucks. Those who could not afford cars found their

Figure 3.19
Construction of the New York subway system, 1901.

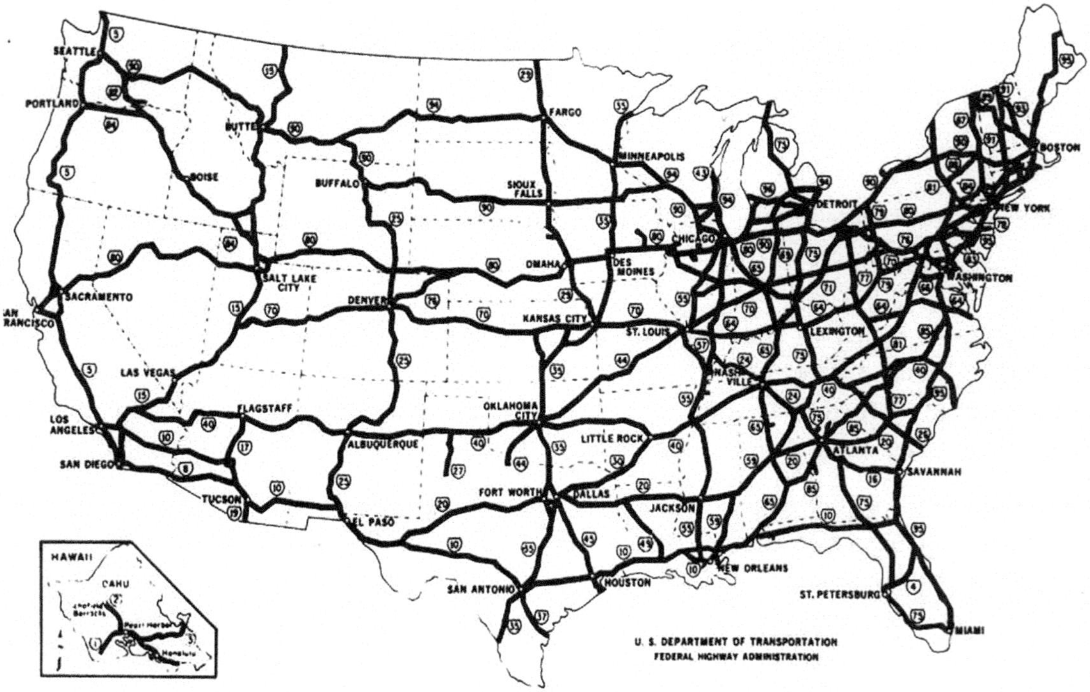

Figure 3.20
US Interstate Highway System. Map lines delineate study areas and do not necessarily depict accepted national boundaries.

mobility options limited. And inner city communities plagued by the evacuation of the middle class saw their tax rolls depleted, impinging on their ability to provide public services.

Today, automobiles are the mainstay of transportation throughout the economically developed world and in many developing countries as well. Even people with lower incomes need a car to function effectively.

3.3.4 The Post-WWII Urban Boom

The three decades following World War II—roughly 1946 to 1978—constituted a unique period in American history. Relatively unscathed by the conflict, unlike its allies and former enemies alike, the United States was unquestionably the largest and most productive economy in the world. It boasted the highest standard of living on the planet. The American steel, automobile, rubber, machine tool, and other industries dominated global markets. Rapid economic growth greatly accelerated social mobility and pulled millions of people out of poverty as the middle class swelled.

These changes were reflected in the dynamic urban landscapes of the country. One of the most important was the enormous growth of the suburbs, which attracted tens of millions of people. By the mid-1960s, more Americans lived in suburbs than in central cities or rural areas; today there are more suburbanites than the other two categories combined. Because race and class in the United States are fused, and Whites typically have higher incomes than do people of color, the exodus to suburbia took the form of White flight.

One factor that contributed to rapid suburbanization was the large expansion of affordable, mass-produced housing. Following the war, and two decades with little residential construction, the demand for housing was enormous. Many returning soldiers, some of whom attended college on the G.I. Bill, were newly married and enjoyed rising disposable incomes. The demand was met in part by the company Levitt and Sons, which pioneered the design and construction of standardized housing at affordable prices. Several suburban communities that grew in the wake of this process were nicknamed Levittowns (Fig. 3.21), sites where only a handful of architectural designs were deployed in great numbers.

The suburbanization of the American population brought with it great economic and social changes. Because retailers follow their customer base, suburban areas began to sprout numerous shopping centers, often filled with chain stores.

However, what was good for suburbs would be disastrous for central cities. White flight eroded their property tax base, reducing the budget for public services such as schools. Declining investment in central city neighborhoods, often leading to decrepit landscapes,

Figure 3.21
Levittown, Pennsylvania.

further drove out the middle class. In short, the growth of suburbs and decline of urban cores reflect capitalism's incessant reproduction of uneven spatial development.

One area that became particularly prominent during this period was the northeastern United States, a region stretching from Boston to Washington, DC that was famously labeled Megalopolis (Fig. 3.22) by the geographer Jean Gottmann. The wealthiest region of the United States, Megalopolis consisted of a well-connected series of metropolitan regions whose expansion hinted they might merge into one giant conglomeration. It includes prominent cities such as Boston, New York, Philadelphia, Baltimore, and the nation's capital, Washington. The commonality of transportation systems and rapidly growing suburbs around each city invited speculation that the area might become one continuous urban area. The emergence of Megalopolis also led planners to rethink urban problems and solutions, envisioning regional rather than local plans.

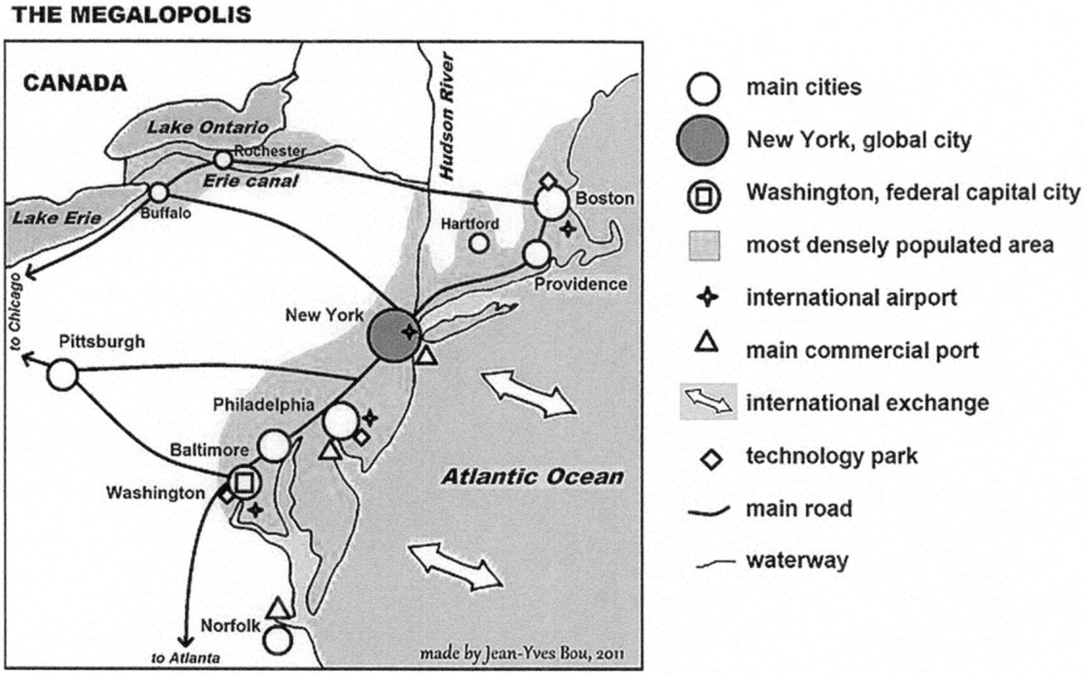

Figure 3.22
Megalopolis. Map lines delineate study areas and do not necessarily depict accepted national boundaries.
(https://commons.wikimedia.org/wiki/File:Megalopolis_nord-am%C3%A9ricaine.jpg.)

3.3.5 Postindustrial American Cities

The postwar boom ended abruptly in the 1970s, when the American and global economies were shaken by the petroshocks, or rapid increases in the price of petroleum. Because oil and gasoline are so central to contemporary ways of life and urban form, the dramatic rise in the cost of energy had enormous impacts on the urban fabric.

First and foremost of these changes was the deindustrialization of vast swaths of the American economy, as automobile, steel, machine tool, rubber, glass, and other types of manufacturing firms began to close plants (Fig. 3.23). It is difficult to overstate the importance and impacts of this process as the United States underwent a rapid change from industrial to postindustrial society. Tens of millions of jobs were lost. The process transformed the once-vibrant Manufacturing Belt into the Rust Belt as capital abandoned a region that had been central to the accumulation of profit for a century. Cities such as Pittsburgh, Detroit, Cleveland, Toledo, Youngstown, Akron, Buffalo, Chicago, Milwaukee, and St. Louis suffered enormous job losses, which in turn caused a drop in real estate values and local retail sales. Urban tax bases suffered accordingly, devastating these cities' efforts to provide public services. Many cities lost large numbers of people as workers fled to the Sunbelt. Because White people had more

Figure 3.23
Closed automobile plant in Detroit.
(https://commons.wikimedia.org/wiki/File:Dilapidated_Packard_Plant,_Detroit,_Michigan_(21560330089).jpg.)

resources to move, many inner city populations became dominated by less wealthy people of color.

Deindustrialization had enormous social as well as economic impacts. Labor unions such as the AFL-CIO and Teamsters suffered declines in membership. Many communities found themselves with large numbers of unemployed workers who suffered from depression, marital problems, and even suicide. Rates of social mobility, especially for people with low levels of education, slowed down enormously.

The flip side of Manufacturing Belt decline was a surge of growth in the Sunbelt. The pattern reflects the uneven spatial development that has always been at the heart of capitalism. Cities in the South and West began to grow rapidly, including Atlanta, Miami, Dallas-Ft. Worth, Phoenix, Los Angeles, and San Diego (Fig. 3.24). Today the majority of the US population

Figure 3.24
Distribution of major US cities. Map lines delineate study areas and do not necessarily depict accepted national boundaries.

lives in the Sunbelt. The growth of this region was driven by several factors, including industrial decentralization, immigration from Latin America, defense spending, and the ubiquity of air conditioning.

This era of rapid transformation was also accompanied by a surge in services of various forms. Services encompass a wide range of industries and occupations, and it makes little sense to speak of "the service sector." As manufacturing declined, both high-end services (e.g., finance, legal services, engineering, advertising, computer software) and low-end ones (e.g., retail trade) grew rapidly. For many people the shift from manufacturing to services involved a decline in real (postinflation) incomes, as low-end services tend to pay poorly.

In the late 20th century, the surge in financial and producer services began to reshape urban areas yet again. Once-derelict downtown areas were reclaimed by corporate America, leading to a new round of growth in skyscrapers. As white-collar professionals began to move to urban cores, many of these places experienced waves of gentrification (Chapter 5). On the urban periphery, many exclave communities witnessed explosive growth.

American cities were also reshaped by the global economy. International trade accelerated, driving growth in ports such as Los Angeles, New York, San Francisco, and Seattle. Foreign investment saw a new round of growth in the automobile industry, this time using robots and just-in-time delivery systems. A considerable proportion of the office buildings in many metropolitan areas is owned by foreign firms.

Simultaneously, large numbers of immigrants entered American cities, primarily from Latin America and Asia. Many southwestern cities saw rapid increases in their populations of Mexican origin, while Miami witnessed an influx of Venezuelans, Nicaraguans, and others. Asian communities, especially on the West Coast, grew rapidly. Often such immigrants form enclave economies (e.g., Little Saigon), much as their predecessors did in the 19th century.

Finally, income inequality has grown markedly under neoliberal capitalism. The rise of numerous high-wage workers associated with producer services, and inadequate construction of new housing, drove up housing prices in many cities. California, in particular, experienced a surge in this regard. The result was only to make housing expensive, and for many people, unaffordable. Some large metro areas began to experience net out-migration as a result. Moreover, high housing costs contributed to the growth of homelessness in many cities (Chapter 6).

In short, given the incessantly restless nature of capitalism, American cities underwent yet another series of changes that reworked the commercial and residential fabric of cities. The landscapes of capitalism are forever changing leading cities to undergo periodic rounds of transformation.

3.4 Chapter Summary

This chapter traced the long and convoluted history of cities in Europe and North America. Their origins began during the medieval era prior to the rise of capitalism, especially during the late feudal era from roughly CE 1100 to 1500. Medieval cities were small, a reflection of the low surplus generated by preindustrial agriculture, crowded, and filthy. The largest rarely exceeded 100,000 people, and most were located in northern Italy, Europe's most prosperous region.

The Industrial Revolution that began in the late 18th century marked an enormous turning point in the history of Western, and global, urbanization. Industrialization was enabled by the harnessing of inanimate sources of energy, which greatly expedited processes of production and transportation. It led to the rise of the factory system, the agglomeration of firms in cities, the industrialization of agriculture, the displacement of large numbers of rural inhabitants, and wave upon wave of rural-to-urban migration. Suddenly cities swelled in size and number.

Industrial cities were highly polluted, congested, and generally unpleasant places to live. Britain led the way in this regard, becoming the world's first industrialized, and urbanized, society. Europe followed suit in the 19th century.

In North America, where several places had hosted Indigenous cities, urbanization unfolded via colonialism. In a maritime world economy, port cities were particularly important. The Spanish set up ports in Puerto Rico and St. Augustine, Florida, and presidios strung like beads along the coast of California. The British erected port cities along the northeastern seaboard, including Boston, New York (originally Dutch), and Philadelphia. With the construction of the Erie Canal, New York cemented its place as the largest city on the continent. For several centuries North America, including what would become the United States, was overwhelmingly rural in nature.

As the Industrial Revolution jumped across the Atlantic Ocean in the early 19th century, American cities began to turn into manufacturing centers. The textile industry flourished in New England, garment production in New York, and heavy industry along the southern shores of the Great Lakes created the Manufacturing Belt, home to some of the country's largest and most productive centers, like Pittsburgh, Buffalo, Cleveland, Akron, Detroit, Chicago, Milwaukee, and St. Louis. The completion of the transcontinental railroads not long after the Civil War ended in 1865 created a national economy, allowing resources from the Midwest and West to be shipped to the industrial centers farther east. Midwestern cities flourished along rivers due to the steamships, and Chicago became the country's second-largest city. The Manufacturing Belt became the largest and most productive region on the planet. On the West Coast, Chinese workers brought in to build the railroads became the nuclei of Chinatowns in Seattle, San Francisco, and Los Angeles.

Urban growth was facilitated by varying forms of urban transportation. During the preindustrial era, most people walked to work. The horse-drawn omnibus was the first form of mass transit. Light rail systems came into being in the late 19th century, including streetcars, and, later, subways. The introduction of the automobile in the early 20th century had massive effects, greatly encouraging suburbanization and urban sprawl.

After World War II, another set of changes swept over the North American urban system. Rising incomes and the baby boom led countless millions, especially White people, to evacuate the cities and head for the suburbs. The construction of the Federal Interstate Highway System starting in 1956 accelerated this process and also led to enhanced ties among cities. In the northeast, the growth of cities from Boston to Washington, DC gave rise to Megalopolis.

The postindustrial cities that characterize the United States today represent the latest urban geography of capitalism. Deindustrialization devastated industrial cities, turning the

Manufacturing Belt into the Rust Belt. As manufacturing jobs disappeared, once-vibrant African American neighborhoods in many cities started to turn into ghettos. Simultaneously, Sunbelt cities in the South and West flourished. Los Angeles became the nation's second largest urban center. The growth of producer services led to renewed investments in urban core areas, revitalizing once-moribund downtown areas and initiating waves of gentrification. Globalization turned some American cities into global cities, like New York, Chicago, and Los Angeles. However, this transformation also brought with it rising inequality, a stressed middle class, and homelessness.

Study Questions

1. Describe the structure of feudal cities. What type of energy did they use?
2. How did the Industrial Revolution change cities?
3. Where in North America did the Spanish initiate urbanization?
4. What forces drove US urbanization?
5. Where is the Erie Canal and what impacts did it have?
6. Summarize the role of immigrants in the American economy and cities in the late 19th century.
7. What were the political machines?
8. What role did the omnibus play in urban transportation?
9. What were Levittowns?
10. Where is the Manufacturing Belt? The Sunbelt?
11. What is Megalopolis?

Further Reading

Warner, The Urban Wilderness: A History of the American City. University of California Press (1995).

J. Williamson, Coping With City Growth During the British Industrial Revolution. Cambridge University Press (2002).

C. Abbott, Thinking about cities: The central tradition in U.S. urban history. Journal of Urban History. 00961442. 22 (6) (1996), 687–701, Available from: http://www.sageltd.co.uk/journals/details/j0187.html, https://doi.org/10.1177/009614429602200602.

W. Cronon, Nature's Metropolis: Chicago and the Great West. (1991). Norton Bradford

J. Bradford De Long, Andrei Shleifer, Princes and merchants: European city growth before the industrial revolution. Journal of Law and Economics. 0022–2186. 36 (2) (1993), 671–702, Available from: https://doi.org/10.1086/467294.

L. Frost, The History of American cities and suburbs: An outsider's view. Journal of Urban History. 00961442. 27 (3) (2001), 362–376, Available from: https://doi.org/10.1177/009614420102700307.

W. James King, T.P. Hughes, Networks of power: Electrification in Western society, 1880–1930. The American Historical Review. 00028762. 89 (2) (1984), 406, Available from: https://doi.org/10.2307/1862567.

J. Foster, Class Struggle and the Industrial Revolution: Early Industrial Capitalism in Three English Towns. Methuen & Co, (2003), 1–341, Available from: http://www.taylorandfrancis.com/books/details/9780203222379, https://doi.org/10.4324/9780203222379.

L. Mumford, The Culture of Cities. (2016).Open Road Media.

Eric H. Monkkonen, America Becomes Urban: The Development of U.S. Cities and Towns, 1780–1980. The American Historical Review. 00028762. 95 (3) (1990), 916, Available from: https://doi.org/10.2307/2164475.
A. Morris, History of Urban Form before the Industrial Revolution. Routledge (2013).
Lewis Mumford, Gideon Sjoberg, The Preindustrial City: Past and Present. American Sociological Review. 00031224. 26 (4) (1961), 656, Available from: https://doi.org/10.2307/2090294.
Gideon Sjoberg, The Preindustrial City. American Journal of Sociology. 0002–9602. 60 (5) (1955), 438–445, Available from: https://doi.org/10.1086/221603.

CHAPTER 4

The Urban Economy

Chapter Outline

Cities are first and foremost machines of production and exchange. Capitalism has been associated with the growth of cities throughout its history. Cities are reflective of the division of labor, the specialization of tasks that makes societies more productive. Thus the organization of cities reflects the organization of production (in contrast to views that begin with individual decision-making). The urban economy is vitally important because it drives the city's growth, processes resources, creates jobs and incomes, allows for consumption, and is foundational to people's opportunities and quality of life.

This chapter examines the urban economy in several steps. It starts by examining agglomeration economies, the benefits of clustering together, which arguably are the primary underpinning of the economic role of cities. Second, it turns to the central business district and its unique role. The third part focuses on the urban export base, the critical, outward-oriented part that is the motor of urban growth. Fourth, it discusses land markets and how different land uses compete for a limited supply of urban space. Fifth, it introduces urban political economy, the webs of power relations that shape urban economies. In the sixth part, it embeds contemporary cities within the wider relations of cognitive-cultural capitalism, the form of knowledge-intensive production that dominates today. Seventh, it examines how telecommunications have reshaped urban space.

4.1 Agglomeration Economies: The Primary Reason Cities Exist

Ever since they originated, the advantage of cities lay in the close proximity that they offered to people, and, later, firms, which greatly reduced transportation costs. These forces are constantly challenged by others that propel the city outward in continuous decentralization.

Urban Geography. DOI: https://doi.org/10.1016/B978-0-443-30281-7.00004-9

A critical reason for the existence of cities, particularly under capitalism, is the insistent tendencies of firms to cluster to take advantage of the network externalities found there. By grouping together, firms minimize transport costs and maximize their access to specialized labor and information. Cities are thus much more than simply conglomerations of people, but consist of networks of interdependencies.

The benefits of clustering together are called agglomeration economies. "Economies" in this sense refers to the gains, such as lower costs and higher profits, that result from clustering. Examples include lower transport costs for inputs and outputs; ability to use specialized services (e.g., repair, consulting, law firms, accounting); access to a specialized labor pool or infrastructure (e.g., fiberoptic lines, airports); and access to specialized information, much of which is transmitted face-to-face. By locating together, firms achieve cost reductions and are hence more profitable than if they located separately. This produces a powerful force that induces firms to locate near one another.

Cities are ideal places for firms to realize agglomeration economies because they minimize the time, cost, and effort of interfirm interactions. For this reason, firms that rely on cities, and particularly centralized locations within them, tend to be those with many linkages to other firms, including purchases of inputs, use of ancillary services, and sales of outputs. Because they reduce the costs of production, agglomeration economies allow larger cities to be more productive and pay higher wages than smaller ones. Agglomeration economies also explain why urban areas are so much more productive than rural ones; the vast bulk of the world's wealth is created in cities.

Agglomeration economies were first theorized by the famous late 19th century British economist Alfred Marshall. He studied the manufacturing districts of cities such as Birmingham, Manchester, and Leeds and noted that "the mysteries of industry become no mysteries; but as it were in the air." Agglomeration economies explained the clustering of jewelry, guns, and garment producers, labor-intensive forms with many linkages to other sectors. Agglomeration economies help to explain the formation of manufacturing complexes such as the Manufacturing Belt and the Ruhr region in Germany. Because of his path-breaking work, centers of production with many agglomerated firms are often called Marshallian districts. Contemporary examples include the clusters of production in northern Italy (sometimes called the "Third Italy"), where firms produce high-quality fabrics, foods, ceramics, glassware, and other goods, and the complex of computer software-related firms in Silicon Valley, California.

In service-oriented cities, agglomeration economies lead firms to engage in considerable face-to-face contact. Although other means of interacting obviously exist (e.g., the telephone, Zoom), face-to-face contact allows for the greatest intimacy, including body language, that is necessary for the creation of trust. Building trust is not easy, but is essential to well-functioning linkages among firms. To facilitate face-to-face contact, firms in cities often prefer downtown locations, where complexes of office buildings allow this process to unfold relatively easily. The attraction of agglomeration is so powerful that firms will pay high rents to be able to take advantage of them. In short, the more accessible a place is, the more valuable the land.

4.2 The Central Business District

At the core of every large city is the Central Business District (CBD), a term coined by the Chicago School of Urbanism in the 1920s (Chapter 6.). Due to its maximum proximity to all parts of the metropolis, this location is geographically advantaged, allowing firms located there the greatest access to urban labor supplies, one another, clients and customers, the infrastructure, and specialized pools of information. Because it has the greatest accessibility, the CBD allows firms there to minimize transport costs and times with other firms. Thus, the CBD offers a comparative advantage in vertically disintegrated types of production in which firms have many linkages to one another, and locating there allows those firms with high transportation costs (i.e., multiple inputs and outputs) to minimize costs by taking advantage of the agglomeration economies readily available in that location. Because accessibility is the major determinant of land values and land use, locations in the CBD typically command very high rents (including the peak value intersection) and are marked by high degrees of vertical real estate development, such as office towers.

The form and function of the CBD have changed significantly over time, reflecting broad structural changes in the local, regional, national, and global economies. In the 19th century United States, when city sizes were relatively small, CBDs were comparatively less well developed. In larger metropolitan regions, they were often characterized by webs of small manufacturing firms and smokestacks. Well into the 1920s, when Burgess and others in the Chicago School first theorized the CBD (Chapter 5), the location was surrounded by blue collar, working class neighborhoods (the "zone of workingmen's homes" in classical social ecology). Many CBDs also contained important retail functions.

During the 1880s and 1890s, however, a period marked by the emergence of producer services and the transition from local to national economies, CBDs became larger and more complex. As multiestablishment corporations began to dominate national economies, CBDs became the command and control centers of cities, including the headquarters of many firms. Their landscapes were thus increasingly given over to skyscrapers, an innovation made possible by technological developments such as structural steel (which allowed for the erection of taller buildings), the elevator, the telephone, and mass transit, which allowed large numbers of workers to be shuttled in and out of downtowns rapidly. The modern CBD would not be possible without large and effective transportation systems, including light rail, subways, buses, and freeways. By the 1960s, this process was more or less complete in the United States, and CBDs were marked by dense complexes of steel and glass towers occupied by workers in producer services such as finance, legal services, accounting, and insurance (Fig. 4.1).

Suburbanization, White flight, and industrial decentralization, however, took their toll. In many American cities plagued by deindustrialization and capital disinvestment, neighborhoods near CBDs in the 1970s and 1980s experienced sharp economic declines, including rising levels of poverty, unemployment, and homelessness. Many downtowns, particularly in the northeast and Midwest, exhibited closed factories and warehouses. The suburbanization of retailing and evacuation of middle class purchasing power led many downtown stores to close.

By the 1990s, as globalization and the explosion of producer services ushered a new round of growth and investments, many CBDs were reclaimed by corporations, a process accompanied by widespread gentrification and associated influx of professional workers. Today, CBDs are typically the primary points of entry for the forces of globalization in the American city: large parts of the downtowns of many US cities are owned by foreign firms, including large real estate interests. Despite the telecommunications revolution, CBDs continue to facilitate face-

Figure 4.1
Downtown areas tend to be dominated by skyscrapers used by producer services, as this example of Sydney, Australia, illustrates.
(https://commons.wikimedia.org/wiki/File:Aerial_View_Chatswood_to_Sydney_CBD_(cropped).jpg.)

to-face interactions, another indication of their long-standing importance to the creation of urban agglomeration effects.

4.3 The Urban Export Base

Because they are always part of a broader capitalist division of labor, cities are always integrated into a wider environment that both reflects and shapes their structure, location, and behavior in time and space. This integration often takes the form of linkages among firms, which buy and sell parts, goods, and services to one another. Put differently, corporate backward and forward linkages are, in part, interregional. Firms thus simultaneously shape and are shaped by their local, regional, and global surroundings. Economic base analysis (sometimes also called export base theory) offers a means of understanding how cities and regions are integrated with one another.

An important part of the geography of capitalism is the tendency for cities and regions to specialize in the production of some outputs and not others (i.e., whether they develop a comparative advantage). The history of American cities offers numerous examples of urban specialization. New York was long known as a garment-producing center and today is the capital of finance; Detroit arose around automobile production; Akron was once the world's capital of rubber manufacturing; Pittsburgh, Buffalo, and Youngstown were major centers of the steel industry; Minneapolis originated as a flour-milling center; Corning, New York, has long been an important glass-producing city; Cincinnati was known as "porkopolis" for its numerous pork-processing firms; Memphis was a cottonseed oil center; Los Angeles is the leading film-producing center of the planet; and Seattle was long closely tied to aerospace through the Boeing Corporation. Similarly, the Great Plains became the breadbasket of North America; Appalachia was a coal-producing area; and the Pacific Northwest exported timber and lumber. Often smaller towns arise as agricultural processing, timber and lumber, or mining centers.

The economic base (or export base) of a country, region, or city is the part of its economy that exports to other places, linking it to markets in other regions and countries. The economic base sells products to a wider market and derives revenues in return. The term "export" in this context does not necessarily mean foreign exports: regions and cities that sell their output to clients in other locations in the same country are effectively exporting their output. For example, Dalton, Georgia, exports carpets to every state in the nation, and New York exports financial and advertising services to clients across the United States. The economic base is the engine that drives a place's economic activity, and is thus vital to the health of a city's or region's economy.

A common but mistaken notion is that cities and regions only export agricultural and manufactured goods. Often, planners view services as only produced and consumed locally.

Yet this idea is mistaken; services are routinely bought and sold interregionally and internationally. Given that the vast majority of the labor force is involved in services, it would be surprising if they were *not* traded among places. The United States derives one-third of its foreign revenues from selling services to other countries. New York sells financial, advertising, and legal services around the world; Pittsburgh, Birmingham, Alabama, and Gainesville, Florida, export health care services to clients in other cities and states; and Los Angeles exports television and film services. University towns export educational services when students move there from other places to attend school. Whenever a radio station in one city sells advertising time to a client located in another, it exports a service. When tourists go to Disney World, New Orleans, Miami, Las Vegas, or San Francisco, those places export services. Even Washington, DC (as well as state capitals) exports government services in the sense that the costs of producing these are paid for by other locations. Thus, countries, regions, and cities certainly can and do have services as an export base.

Economic base theory divides the local economy into two broad sectors, the economic base, or basic sector (B), and the nonbasic sector (NB). The basic sector is export-oriented (selling agricultural or manufactured goods or services); it is the motor of the urban economy. When export bases are healthy, they generate jobs, incomes, and attract immigrants, contributing to population growth. In contrast, the nonbasic sector recycles corporate and personal incomes. Nonbasic economic sectors include retail trade, eating and drinking establishments, real estate, and personal services (e.g., hair salons, lawyers, repair shops). Essentially, these functions cater to local demand. From the perspective of economic base analysis, therefore, the nonbasic sector occupies a less important role in shaping the local economy than does the economic base. Of course, the division between these two sectors is not always so simple. Some cities may export retail services when people go there to shop, and some sectors, such as legal services, may be both basic and nonbasic (e.g., a law firm that handles both corporate mergers in another city and local divorces).

We can understand these relations using a simple economic base model, in which total employment (T) equals the sum of the basic (B) and nonbasic (NB) sectors, or

$$T = B + NB.$$

We now introduce the concept of a multiplier, m. The multiplier reflects the degree to which the basic sector drives the total economy. There are several types of multipliers and different ways of estimating them, but here we define the multiplier (m) as the ratio of total to basic sector employment, or

$$m = T/B.$$

Substituting (B + NB) for T in the preceding equation yields

$$m = (B + NB)/B = 1 + NB/B.$$

This approach allows us to estimate the degree to which changes in the basic sector cause changes in the total economy (ΔT), that is,

$$\Delta T = m\Delta B$$

Consider a simple example. Say that a small town has 1000 people in the labor force. (Total population will be higher, depending on the labor force participation rate.) Of that 1000, 400 work for a local textile mill. If the mill hires an additional 100 people, what will be the impact on the community? We can answer this question simply enough by defining B=400; therefore, NB=600, Thus, the textile mill's multiplier is

$$M = 1 + 600/400 = 2.5.$$

The change in total employment that an additional 100 in the basic sector generates is thus

$$\Delta T = m\Delta B = 2.5.100 = 250.$$

The additional 100 jobs in the textile mill created 250 jobs in the community (including the 100 new jobs in the basic sector). Because the multiplier is always greater than zero, changes in the basic sector always create larger changes in the rest of the local economy; that is, since $m > 0$, $\Delta T > \Delta B$

How did this process operate? The economic base model holds that when the basic sector expands (or contracts), those changes reverberate through the local economy in a series of interfirm linkages as well as changes in consumer spending. The total changes can be decomposed into three constituent parts: direct, indirect, and induced effects. When the economic base changes (increasing or decreasing employment), it induces direct effects; second, there are changes in employment in firms that sell goods and services to the export base firm(s) through subcontracts, or indirect effects. For example, when the textile mill in the preceding example expands, it may increase its purchases of equipment, repair and maintenance services, office machinery, advertising services, legal services, and other inputs. Thus, the size of a multiplier—and thus the impacts of the basic sector on the rest of the urban economy—is closely tied to the linkages among firms. Every firm purchases inputs from other firms, that is, every purchase is a sale, depending on whether one takes the perspective of the buyer or the seller. Firms that supply the textile mill with inputs of goods and services will enjoy increases in sales as their client buys more inputs from them. These are the indirect effects.

The size of the multiplier in many ways reflects the spatial distribution of backward linkages from the basic sector, that is, the geography of subcontracts. To the degree that the basic sector firm subcontracts with suppliers located far away, many of the benefits of its growth will be exported as well as "leakages." Conversely, the more the firms in the basic sector subcontract locally, the higher the local multiplier effect will be. Sometimes governments require that public sector subcontracts (e.g., for construction) only use local firms for this reason.

Finally, the third part of total changes resulting from multiplier effects are changes in consumer spending resulting from changes in the basic sector. When the exporting firm and its subcontractors hire more workers, there is more disposable income circulating in the local economy. These expenditures, or induced effects, increase jobs in the nonbasic part of the economy. Thus, higher wages will lead to higher expenditures on real estate, food, entertainment, transportation, and other consumer goods and services. The size of induced effects will depend on the number of workers affected, their salaries, and their spending habits and preferences.

It is important to note that multipliers are double-edged swords, that is, they may work against as well as for a region. Changes in the basic sector can be negative as well as positive, such as when a steel plant or auto assembly plant shuts down. Thus, if $\Delta B < 0$, then $m\,\Delta B$ will be <0. For example, if a textile mill closes down, or if an automobile factory reduces its labor force (perhaps as part of a strategy of industrial reorganization and relocation overseas), then the multiplier effects will take hold, amplifying the initial losses. Subcontracts to supplying firms dry up, reducing the indirect effects. Lower workers' incomes in turn lead to negative induced effects, causing local businesses to suffer and real estate prices to decline. For this reason, the downtowns of many deindustrialized cities often have numerous storefronts that have been closed and housing that is relatively cheap.

Finally, the diversity of the local economy is important. Typically, the basic sector declines and the nonbasic sector rises as cities increase in population. This pattern reflects the fact that large cities usually have a diversified economic base. The Los Angeles region, for example, has an economic base in the film industry as well as finance, software, the port, and aerospace. Conversely, small towns often have very narrow economic bases centered on one industry, and often only one firm, such as agricultural processing centers, mining towns, and lumber-dependent cities.

4.4 Urban Land Markets

The structure of land use plays an enormous role in urban economies. Land use simultaneously reflects and shapes the patterns of urban activity. Land uses are not randomly distributed throughout the city; they are products of systems of production and social reproduction, as well as the state (Chapter 6).

The first attempt to theorize how capitalism shapes land use was made by a wealthy German landlord in the early 19th century, Johann Heinrich von Thünen. Observing patterns of land use on his estates, he derived a mathematical model to express the relations between prices, transport costs, and the rate of profit on a given parcel of land (rent). Specifically, he noted that

$$R = Y(p-c) - Ytd,$$

where

R = land rent; Y = yield per unit of land;

p = market price per unit of commodity;

c = production cost per unit of commodity;

t = transport cost per unit of distance; and

d = distance to market.

Graphing this relation yields a bid-rent line, a fundamental idea in urban economics. A bid-rent line reflects how much a user of land is willing to pay for it at different locations. Note that this expresses how the rate of profit declines with distance from the market (Fig. 4.2). At distance 20, the rate of profit is zero, meaning that at that point and farther it is not profitable to have *any* kind of land use, marking the outer edge of the city.

The precise shape of the line will be dictated by the variables p (market price), c (production costs), and t (transport costs). Note that the rent, or profitability per unit area, drops with distance from the market as rising transport costs eat up profits. *This is a fundamental concept in urban economics: the more accessible an area is, the higher the profitability.* Conversely, less accessible areas farther from the market have lower rates of profitability. (In reality, these bid-rent lines are more likely to be curves, but we'll keep things simple here).

Bid-rent lines and curves are not static, but fluctuate as market prices, production costs, and transport costs change. These changes reflect the dynamism of capitalism and the ways in which new production systems are inscribed into the urban landscape.

Von Thunen noted that different land uses, with different yields, prices, and production and transport costs, competed for the same land. As a result, there are multiple bid-rent curves floating over a given area. In a capitalist economy, the bid-rent line that generates the highest rate of profit in a particular place will be the one that decides the land use there. Fig. 4.3 illustrates multiple bid-rent

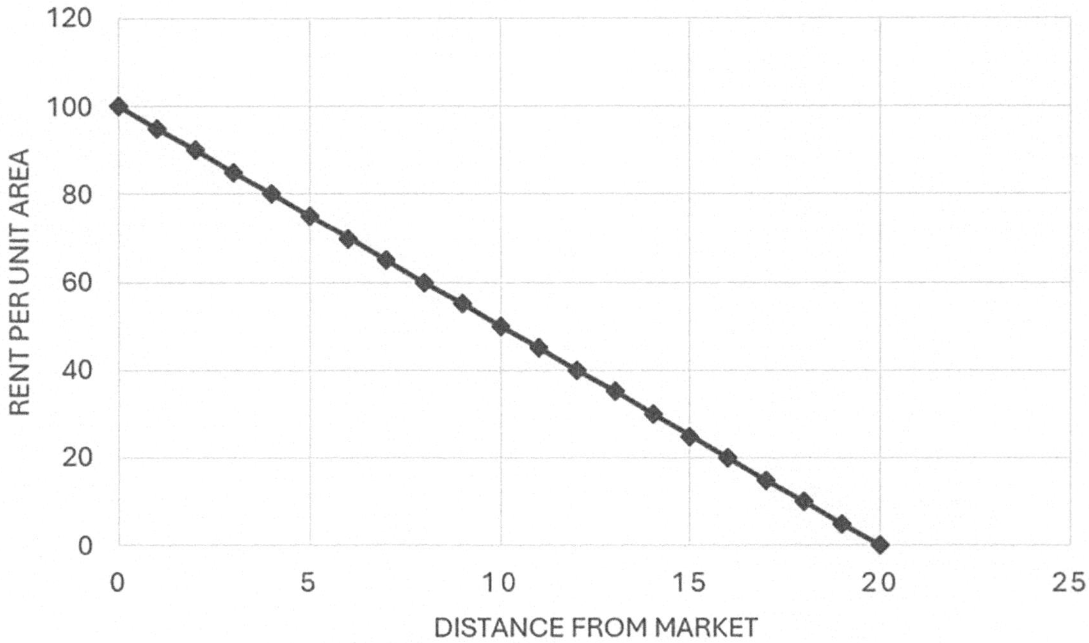

Figure 4.2
A simple bid-rent line.
(From Author.)

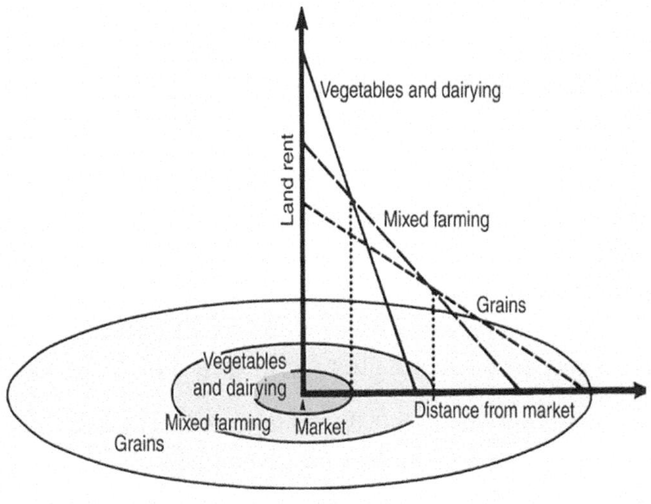

Figure 4.3
The von Thünen model with multiple land uses and expanded to two dimensions.
(From Vector image based on Figure 10-52 in James M. Rubenstein's textbook The Cultural Landscape (11th Edition, 2014), recreated by the author and released into public domain. https://commons.wikimedia.org/wiki/File:Von_Thunen_Model.svg.)

lines and the profit-maximizing land use pattern that results. Applying this model to two dimensions rather than a simple line yields concentric rings of land use. In short, controlling for all else, in a capitalist economy competing land uses will create rings of the same land use centered on the city center.

As a tool to explain agricultural land use, the von Thünen model is not particularly effective, given the globalization of food production, and is rarely used in that way. But it nonetheless retains some utility in explaining urban spatial structure. Thus, as illustrated in (Fig. 4.4), bid-rent lines for offices, retail trade, industry, and residential purposes segregate land use at ever-farther distances from the city. Firms that need the agglomeration economies of the city center will outcompete other land uses to locate there, generating the highest rate of profit. Conversely, more land-extensive uses, such as warehousing or suburban developments, locate farther from the city center where land is cheaper. In short, the observed pattern of land uses is the one that maximizes rents, barring distortions from the natural environment (e.g., proximity to water) and government actions, which are not decided by profit-maximization.

This simple approach to urban land markets helps us to understand why the city appears the way it does. For example, it explains why there are complexes of offices in the CBD; why there is relatively little residential land use near downtown; and why outlying areas are given over to low-level and relatively low-density types of functions such as housing.

4.5 Urban Political Economy

The economies of cities are not simply the result of markets, powerful and important though they are. Urban space is also deeply political in nature, the result of social forces such as class relations that structure production under capitalism. Indeed, class conflict has played out regularly over space, from strikes at the workplace to struggles over land use. Politics is about power: who has it, who gets or loses it, and how different parties seek to achieve their ends. A political economy perspective unites the economic and political dimensions of the city. It is thus vital to see urban areas as centers of political economy, in which political and economic interactions are always fused. The power relations that infuse cities take a variety of forms.

Capitalism is not simply a bunch of individuals pursuing their own ends. It is also riven by class divisions and contradictions. Class relations structure most aspects of cities, largely as the two major classes—workers and capitalists—pursue mutually exclusive goals. Conflicts between labor and capitalists have rocked cities since the capitalist city began. These take the form of strikes, demonstrations, and crackdowns on dissent. During the 19th century, this process saw garment workers (mostly women) going on strike for higher wages and better working conditions (Fig. 4.4); at times, corporate thugs such as the Pinkertons were called out to put down strikes by force. At times general strikes saw almost all workers walk off the job. In the 1930s, American cities were rocked by enormous strikes led by the steel and

Figure 4.4
Striking garment workers in New York City, 1913.
(https://commons.wikimedia.org/wiki/File:Garment_Workers_on_Strike,_New_York_City_circa_1913.jpg.)

automobile unions, which resulted in the two-day weekend and paid overtime. Today, class conflict may take the form of hospital or retail workers striking for much the same reason.

A striking example of urban political economy is the ability of firms to use place against labor. In general, capital is more mobile than labor, for workers are sutured to places through their everyday lives, family and friendship networks, and their reliance on local services. For this reason, firms frequently threaten to leave a town or city in order to obtain concessions from workers or the local state. Often this involves demands for exemptions from local property, income, or sales taxes. Firms with few connections to a community are most apt to engage in this type of behavior, notably those with few local investments and many distant potential substitutes. Examples include light manufacturing firms. Capital is usually more mobile than labor; workers, after all, have dense networks of social relations, including family and friends. Firms may threaten to leave to wring concessions from municipal governments, such as tax breaks or demands for training programs or infrastructure changes (e.g., new sewer lines).

But not all types of capital are equally mobile. Some companies have deep roots and extensive linkages to their communities, such as utilities, local newspapers, real estate developers, law firms, and retailers. They cannot simply pick up and leave, and there are few if any other places that could substitute. They are essentially trapped by the immobility of their investments in the local environment. This situation leads them to be invested in the health and future of their communities. The most mobile forms tend to have minimal capital investments and ready access to distant markets and suppliers, a mobility greatly enhanced by globalization and ever-increasingly long supply chains; for these firms, one place is pretty much like any other. However, some firms such as utility companies, law firms, real estate developers, small stores, or local newspapers are, like workers, deeply embedded in places, reliant on local customers and supply networks. For this latter group of firms, places are not easily substitutable for one another.

Place-bound firms often form *growth coalitions* to maximize their political influence. Growth coalitions are typically organized through chambers of commerce, which are alliances or networks of businesses that serve as spokespersons for the business community. Firms contribute dues and hire an executive to run it. Chambers of Commerce typically have substantial clout over urban governance. They generate a large share of the campaign contributions that elect municipal employees, and may use lobbyists after the election to sway policy makers. In return, they shape local public policy, notably zoning codes and environmental regulations. Growth coalitions consist of firms and other actors that are essentially trapped in place by their local dependence and investments. Examples include local developers, retailers, and utility companies. The goal of growth coalitions is to attract capital, increase property values, and create a "good business environment," which often entails low corporate taxes. Driven by the ideology that growth is the same as progress, they may vehemently oppose any public initiative that curtails growth, such as zoning codes or environmental regulations. Growth here is seen as synonymous with progress. Growth coalitions strenuously oppose any limitations on growth, such as zoning ordinances that they regard as unfair. Firms are often well organized in coalitions that give them maximum political effectiveness. This gives them great leverage over local governments and an ability to shape local policy, such as zoning.

Growth coalitions often engage in boosterism to promote their city's economy, including advertising, to attract firms, visitors, and money. Such efforts are called place promotion, and appear in business, marketing, and tourism media channels. Often cities invent slogans to attract visitors. A famous example is the "I ♥ New York" campaign (Fig. 4.5). Growth coalitions often come up with slogans meant to project a vibrant local culture and economy. Chicago adopted "at the heart of everything." Dallas went with "the city of choice for business." Baltimore's is "more service, more choices." And in Milwaukee, "the city that works for your business." Chicago's growth coalition came up with "the center of everything"; Norfolk, Virginia, has "where business is a pleasure"; and Troy, New York, invented "a great location for your

Figure 4.5
The famous "I ♥ New York" logo, a spectacularly successful marketing campaign.

company." This type of place promotion has a long history. In the 19th century, cities jockeyed with one another to have canals and railroads pass nearby, hoping to benefit from the traffic. In the 20th, they did the same with highways. With the rising importance of tourism, some places tout their community as a great place to visit or live, focusing on drawing potential visitors: hence Memphis, Tennessee, argues it is the place "where the music never stops, and most of it started." Attracting visitors, tourists, and residents may lead cities to invest in concert halls, museums, sports stadia, and convention centers.

Another dimension of urban political economy concerns the competition among cities. As the economies of most Western nations began to slow down in the late 20th century, competition among cities for jobs began to resemble a zero-sum game, in which one city's gain was another's loss. The withdrawal of the national government from many functions increasingly left cities to their own devices. This may involve the use of slogans that emphasize their competitiveness (hence Indianapolis floated "move over New York, apple is our middle name."). Unfortunately, this competition, often manifested through tax exemptions and subsidies for firms, can resemble a "race to the bottom," in which cities auction themselves off to the highest bidder. This growth strategy can change the nature of urban planning, replacing traditional priorities such as public services or protecting the environment with a win-at-all costs worldview in which attracting firms and jobs matters more than anything else. To do so, firms may change zoning ordinances, offer subsidies or infrastructure (e.g., electrical and sewer lines), training programs, low-interest loans, and so forth. Prominent examples include the desperate attempts by many cities to woo Amazon when it announced a search for a second headquarters, or attempts by cities and counties to lure makers of electric vehicle parts today. Ironically, the more vigorous the competition among cities, the greater their subordination to capital becomes.

4.6 Cities and Cognitive-Cultural Capitalism

Since the late 20th century, the hallmark of the global economy is what is often called "cognitive-cultural capitalism," a particular form of information-intensive capitalism that relies heavily on the commodification of culture. In such a labor market, employers place a premium on laborers with the intellectual capacities to process, analyze, and communicate large volumes of information. Firms engaged in such activities rely heavily on the agglomeration economies of large cities. The rise of cognitive-cultural capitalism is thus manifested in the explosive growth of producer services across the globe, including finance, legal services, computer software firms, advertising, public relations, engineering and architecture, and other assorted companies that engage in the production of specialized expertise. It also includes the management functions of many large corporations, as well as research and development. Such jobs often require technical experience and involve decision-making in a variety of complex tasks, which are difficult to automate. Historically, intellectual labor has been marginal to the production process; under cognitive-cultural capitalism, it has moved to center stage.

The emergence of cognitive-cultural capitalism is synonymous with the growth of global neoliberalism. From global cities that act as the command-and-control function of the world economy to smaller urban areas, these places have increasingly developed a competitive niche as Marshallian nodes embedded within worldwide commodity chains. Most famously, Richard Florida's celebrated notion of the creative class has drawn enormous attention. The creative class involves an array of occupations and workers involved in information processing, notably software, finance, producer services, parts of health care, law, the media, the arts, entertainment, and education. They are clustered in metropolitan regions where urban agglomeration economies are powerful magnets that link firms in dense networks of interactions. The creative class reflects the convergence of the cultural and the economic, of symbolic and use values. Such workers tend to like their jobs and put a premium on self-expression. Their work and personal lives are deeply intertwined. The creative class is disproportionately represented in startups, and its education and creativity are essential to invention, innovation, solving problems, product development, marketing, and entrepreneurship. In this reading, the skilled professionals so central to cognitive-cultural urban economies place a high value on amenities, recreational opportunities, and local social climates that value diversity, including ethnic differences and tolerance of alternative lifestyles (notably LGBTQIA+ communities). Cities that are tolerant, Florida maintains, are economically healthy; those that are restrictive and xenophobic are unlikely to attract the talent they need to remain competitive. Regions hoping to cultivate high technology sectors in particular must foster political contexts marked by social and cultural tolerance. In addition, it must be noted that the creative class—essentially yuppies—is sustained by vast legions of low-paid service workers who are disproportionately women, immigrants, and ethnic minorities.

Once thought to be confined to the cities of the developed West, the creative class has also expanded into many cities throughout the world, notably in Europe and East Asia. In Asia, cities such as Tokyo, Seoul, Shanghai, Hong Kong, and Singapore exhibit large numbers of workers in knowledge-intensive occupations.

4.7 Telecommunications and the City

The microelectronics revolution has major impacts not only on everyday life but also on urban form. Notably, the introduction of digitized telecommunications systems greatly altered urban form and labor markets.

There exists considerable confusion about the real and potential impacts of telecommunications on urban structure, in part due to the long history of exaggerated claims made in the past, particularly by those subscribing to "postindustrial" theory. Often such views hinge upon a simplistic, utopian technological determinism that ignores the complex, often contradictory, relations between telecommunications and local economic, social, and political circumstances.

For example, repeated proclamations that information systems would allow everyone to work at home via telecommuting, dispersing all functions and spelling the obsolescence of cities, have fallen flat in the face of the persistence of growth in dense urbanized places. In fact, advances in videoconferencing notwithstanding, telecommunications are generally a poor substitute for face-to-face meetings, the medium through which most sensitive corporate interactions occur, particularly when the information involved is irregular and unstandardized in nature. Most managers spend the bulk of their working time engaged in face-to-face contact, and no electronic technology can allow for the subtlety and nuances critical to such encounters. For this reason, a century of technological change, from the telephone to fiber optics, has left most high-value-added, white collar, administrative command and control functions clustered in downtown areas (despite their high rents), their agglomerative advantages intact, if not unchallenged. In contrast, telecommunications are ideally suited for the transmission of routinized, standardized forms of data such as back office clerical functions. In short, contrary to much received opinion, there is no a priori reason to believe that telecommunications inevitably lead to the dispersal or deconcentration of functions; by allowing the decentralization of routinized ones, information technology may enhance the comparative advantage of inner cities (albeit with jobs generally filled by suburban commuters). Information systems thus allow for the simultaneous concentration and deconcentration of economic activities.

Few domains of economic activity have been as heavily affected by telecommunications as low-level, routinized back office functions. Back offices perform clerical functions such as data entry of office records, telephone books, or library catalogs, stock transfers, payroll, or billing information, bank checks, insurance claims, and magazine subscriptions. These tasks involve unskilled or semiskilled labor, primarily women, and frequently operate on a 24-hour-per-day basis. Back offices have few of the interfirm linkages associated with headquarters activities and require extensive data processing facilities, reliable sources of electricity, and sophisticated telecommunications networks.

Historically, back offices have located adjacent to headquarters activities in downtown areas to ensure close management supervision and rapid turnaround of information. However, under the impetus of rising central city rents and shortages of sufficiently qualified (i.e., computer literate) labor, many service firms began to uncouple their headquarters and back office functions, moving the latter out of the downtown to cheaper locations on the urban periphery. Most back office relocations, therefore, have been to suburbs.

Given the increasing locational flexibility afforded by the Internet, back offices have begun to relocate on a much broader, continental scale. Under the impetus of new telecommunications systems, many clerical tasks have become increasingly footloose and susceptible to spatial variations in production costs. New digital call distribution systems have made possible the relocation of phone services that were once confined to centralized locations. Internationally, this trend has taken the form of the offshore office. The primary motivation for offshore relocation is low labor costs, although other considerations include worker productivity, skills, turnover, and benefits. Offshore offices are established not to serve foreign markets, but to generate cost savings for US firms by tapping labor pools in developing countries where wages are as low as one-fifth of those in the United States. Notably, many firms with offshore back offices are in industries facing strong competitive pressures to enhance productivity, including insurance, publishing, and airlines. Offshore back office operations remained insignificant until transoceanic fiberoptics lines made possible geographic flexibility on an international scale. Such functions may be either subsidiaries of multinational firms or they may operate under contract with US-based businesses. Inputs are sent by air or emailed to offshore processing facilities. After processing, generally a few days at most, the results are returned via satellite or dedicated telephone or fiberoptic line. Thus, the capital investments in such operations are minimal and they possess great flexibility, maximizing their ability to choose among locations based on slight variations in cost or profitability.

Several New York-based life insurance companies have erected back office facilities in Ireland, with the active encouragement of the Irish Development Authority. Often situated near Shannon Airport, they move documents in by Federal Express and the final product back via satellite or one of the numerous fiberoptic lines that connect New York and London. Likewise, the Caribbean, particularly Anglophone countries such as Jamaica and Barbados, has become a particularly important locus for American back offices. American Airlines paved the way in the Caribbean when it moved its data processing center from Tulsa to Barbados; through its subsidiary Caribbean Data Services (CDS), it expanded operations to Montego Bay, Jamaica and Santo Domingo, Dominican Republic. In Asia, Manila, Philippines, has emerged as a back office center for British firms, with wages 20 percent of those in the United Kingdom. Such trends indicate that telecommunications may accelerate the offshoring of many low-wage, low value-added jobs from the United States, with dire consequences for unskilled workers.

Figure 4.6
Call center.
(https://commons.wikimedia.org/wiki/File:Callcentre.jpg.)

A similar form of low-wage, low-value-added services involves call centers. Call center functions include telemarketing, customer assistance, and phone orders, often with designated 1–800 numbers. They range greatly in size, from as few as five to as many as several thousand employees. Like back offices, call centers are primarily screen-based and do not require proximity to clients. The major cost consideration is labor, and the workforce consists primarily of women with lower skills, and high turnover rates are common. They are thus the epitome of a footloose industry, disembedded from their local milieu and highly mobile. There are an estimated 100,000 call centers within the United States (Fig. 4.6), which employ between three and five percent of the national labor force, the majority of which are located in urban or suburban locations. Like back offices, call centers have become increasingly globalized. India, for example, has attracted a significant number of customer service centers near its software capital of Bangalore, where workers are trained to speak American English and are able to gossip with customers about pop culture.

One increasingly important impact is "telework" or "telecommuting," in which workers substitute some or all of their working day at a remote location (almost always home) for time usually spent at the office. (The self-employed do not count as teleworkers because they do not substitute it for commuting.) Telework is most appropriate for jobs involving mobile activities or routine information handling such as data entry or directory assistance. Proponents of telework claim that it enhances productivity and morale, reduces employee turnover and office space, and leads to reductions in traffic congestion and accidents (especially at peak hour), air pollution, and energy use. Roughly two million people in the United States (1.5% of the national labor force) telecommute one to two days per week (Fig. 4.7). The COVID-19 pandemic led large numbers of workers to work from home, often conducting meetings over platforms such as Zoom. Further growth of telecommuting will encourage more decentralization of economic activity in suburban areas.

Figure 4.7
Telework has given many workers the ability to work at home.
(https://commons.wikimedia.org/wiki/File:USMC-100324-M-6847A-001.jpg.)

However, there are reasons why telecommuting may *increase* the demand for transportation rather than decrease it. Although telecommuters spend fewer days at their workplace, they tend to live farther from their workplaces than do daily commuters, and thus their total *weekly* as opposed to daily commutes may actually rise. Moreover, the time freed from commuting may be spent traveling for entertainment or social purposes. Thus, telecommuting may alter the reasons for travel, but not necessarily the frequency or volume of trips. Finally, by reducing traffic congestion during commuting hours, telecommuting may generate induced effects, whereby other drivers formerly inhibited from taking to the road may be induced to do so. In short, the trade-off between telecommunications and commuting is far from clear.

Another impact of information systems on urban form is transportation informatics, which includes measures such as smart metering, electronic road pricing, synchronized traffic lights, automated toll payments and turnpikes, travel advisory systems, electronic tourist guides, remote traffic monitoring and displays, and computerized traffic management and control systems, all of which are designed to minimize congestion and optimize traffic flow (particularly at peak hours), enhancing the efficiency, reliability, and attractiveness of driving. Yet other impacts include closed-circuit television monitoring of urban public space (transportation corridors and rapid transit), which has become increasingly ubiquitous and has raised concerns over invasion of privacy.

4.8 Chapter Summary

Cities under capitalism are, at their core, vast machines for the production of commodities, in the process creating labor markets, generating profits, and structuring urban space around the logic of firms' locational needs. Cities are obviously much more than this, including spaces of social reproduction (Chapter 6), but the urban economy is the driving force that dictates a city's prosperity and economic health.

The major reason behind the formation of cities is agglomeration economies, the benefits to be derived from clustering together. By locating near one another, firms save on transport costs of inputs and outputs, create specialized labor markets, and have access to information, often by face-to-face contact and word of mouth. Locational proximity means that firms are more productive, and profitable, when they are situated near each other than if they are located far apart. Agglomeration economies are most visible in the Central Business District (CBD), the core of the city that is often held to be "the tail that wags the dog."

Because cities are always locked into networks of places (Chapter 5), many specialize in the production of a few goods or services. To realize profits, firms sell their outputs to customers and clients in other cities (and, increasingly, around the world). The part of the urban economy that sells its output elsewhere, and thus derives nonlocal revenues, is called the export base, which is the engine of economic growth. Export bases are tied to the rest of the urban economy through complex webs of backward linkages and suppliers. Each change in the export base reverberates through the city in a series of multiplier effects.

Because agglomeration economies render centralized locations to be the most advantageous, firms pay a premium to obtain them in the form of high rents. The land markets of cities are structured around high rent areas in the urban core, as witnessed by the proliferation of skyscrapers downtown, and declining rents with distance away from the CBD. Different land uses compete with one another, and landowners rent or sell land to firms that will generate the highest rate of profit. For this reason, we tend to see offices downtown, and manufacturing and housing farther away, with space-extensive land uses like agriculture confined to the urban periphery. In short, the structure of urban land use reflects the structure of commodity production.

Urban economies are not simply "economic" in nature but are also deeply political. The advantage of a political economy perspective is that it unites the political and economic dimensions of urban life. Firms in cities organize and cooperate when it is in their shared interests. Often they form growth coalitions that work to influence local governments and promote their cities to investors and tourists.

Capitalism constantly changes over time, assuming very different forms in varying historical moments. Starting in the late 20th century, global capitalism underwent a dramatic transformation, becoming much more information-intensive in nature. One label for this new form is cognitive-cultural capitalism. Not coincidentally, this occurred in tandem with the microelectronics revolution and waves of automation of blue-collar jobs. The fastest growing parts of the urban economy have been centered on producer services, skilled, white-collar jobs in finance and related business services, which demand well-educated workers and pay them high salaries. A famous theory about this transformation is that of the creative class, which focuses on the role of such workers in driving various facets of urban space, such as recreational spending and gentrification.

Finally, as cognitive-cultural capitalism took hold, telecommunications came to play increasingly important roles in restructuring urban space. There is a long history of this process going back to the telegraph and telephone. But the microelectronics revolution and the digitization of information gave rise to high-capacity information systems such as fiberoptics and the Internet. The introduction of these systems expedited the movement of many pink-collar jobs in back offices, such as data entry and call centers, to the urban periphery and even beyond. Business meetings over the Internet (e.g., Zoom) also reflect how digital systems have become embedded in urban space. Telecommunications threaten the agglomeration economies that underlie cities' reasons for existence. As a result, jobs that require face-to-face contacts and social skills, and thus are not easily performed via computer-mediated technologies, became more important.

Study Questions

1. What are agglomeration economies and why are they so important to cities?
2. Describe the role of the CBD in the urban economy.
3. How does the export base drive urban economic growth?
4. What is the von Thünen model?
5. Why is land near the city center more expensive than that on the periphery?
6. How do competitive rents maximize returns to landowners at different distances from the city center?
7. Give an example of urban class conflict.
8. How do firms use places against labor?
9. What is a growth coalition and what does it do?
10. How did cognitive-cultural capitalism change urban labor markets?
11. How did telecommunications reshape the geography of back offices?

Further Reading

N. Aveline-Dubach, Globalization and Dynamics of Urban Production Globalization and Dynamics of Urban Production. wiley, France wiley, France (2024), 1–247, Available from: https://www.wiley.com/en-in/Globalization+and+Dynamics+of+Urban+Production-p-9781394257485.

David A. Banks, The City Authentic: How the Attention Economy Builds Urban America University of California Press. University of California Press (2023). Available from: https://doi.org/10.1525/9780520383470.

R. Florida, The Rise of the Creative Class. Routledge (2004).

J. Logan, H. Molotch, Urban Fortunes: The Political Economy of Place. University of California Press (2007).

Richard Florida, Cities and the Creative Class. Informa UK Limited, Informa UK Limited (2005), 35–54, Available from: https://doi.org/10.4324/9780203997673-3.

C. Jones, Urban economy: Real estate economics and public policy Urban Economy: Real Estate Economics and Public Policy. Taylor and Francis, United Kingdom Taylor and Francis, United Kingdom (2021), 1–316, Available from: https://www.taylorfrancis.com/books/9781000465785. https://doi.org/10.1201/9781003027515.

Judd, City Politics (10th ed.) Routledge. (2018).

N. Kleniewski, A.R. Thomas, Cities, change, and conflict: A Political economy of urban life: Fifth edition Cities, Change, and Conflict: A Political Economy of Urban Life: Fifth Edition. Taylor and Francis Inc., United States Taylor and Francis Inc., United States (2019), 1–384, Available from: https://www.routledge.com/products/9781138604476. https://doi.org/10.4324/9780429022920.

A. Scott, A world in emergence: Notes toward a resynthesis of urban-economic geography for the 21st century. Urban Geography. 02723638. 32 (6) (2011), 845–870, Available from: https://doi.org/10.2747/0272–3638.32.6.845.

Allen John Scott, Beyond the creative city: Cognitive–cultural capitalism and the new urbanism. Regional Studies. 0034–3404. 48 (4) (2014), 565–578, Available from: https://doi.org/10.1080/00343404.2014.891010.

Michael Storper, Thomas Kemeny, Naji Makarem, Taner Osman, The Rise and Fall of Urban Economies: Lessons from San Francisco and Los Angeles. Stanford University Press, Stanford University Press (2015). Available from: https://doi.org/10.11126/stanford/9780804789400.001.0001.

Wyly Elvin, The city of cognitive–cultural capitalism. City. 1360–4813. 17 (3) (2013), 387–394, Available from: https://doi.org/10.1080/13604813.2013.807014.

CHAPTER 5

City-Systems

Chapter Outline

Cities never exist in isolation but are always entwined with one another in complex webs of interactions, including trade, migration, and flows of capital and information. Thus cities are systems situated within systems of cities.

This chapter examines city-systems in several ways. It starts with an overview of urban hierarchies, the nested sets of cities that define national urban systems. The second part discusses small towns, the layer of places at the lowest end of urban systems. Third, it summarizes early approaches such as Central Place Theory and the rank-size rule, as well as primate cities. Fourth, it addresses interurban flows, the movements of people, goods, capital, and information that lock cities together. In the fifth part, it describes city-regions, in which cities are linked to surrounding towns and rural areas. Finally, it explores the vital question of global cities, a small set of large metropolitan areas that act as command-and-control centers of the world economy.

5.1 Urban Hierarchies

Because cities of varying sizes have different economies, networks of cities are called urban hierarchies, ranging from small hamlets to enormous metropolitan areas. Typically they are dominated by a single, or at times a handful, of large cities, and include progressively more centers with small sizes; there is an inverse relationship between size and number of cities. Often cities within hierarchies are classified as being part of different tiers of urban areas.

Urban Geography. DOI: https://doi.org/10.1016/B978-0-443-30281-7.00005-0

While urban hierarchies existed in a fashion before capitalism, they were poorly developed, with relatively few connections among cities. The growth of capitalism in the 16th and 17th centuries, however, led to the rapid emergence of urban networks that stretched over increasingly long distances. Urban hierarchies reflect how the dynamics of capitalism unfold across interurban space. These are systems of cities interlinked with one another to varying degrees, and range from enormous, globalized metropolitan areas to small towns with rural hinterlands. Moreover, urban hierarchies reflect complex webs of interdependence, meaning that changes in one city inevitably radiate out to affect others. Flows of people, goods, and information are part of the supply chains that connect cities in webs of market relations. Thus, urban hierarchies reveal how cities are positioned relationally rather than simply located in absolute space.

The importance of cities within the hierarchy is not simply a matter of population size, but rather of the local economic engines of urban areas and their respective roles in the national and global division of labor. Put differently, urban hierarchies are the urban manifestation of the capitalist spatial division of labor. This allows cities to climb the hierarchy at times. For example, in the late 20th century, Charlotte, North Carolina, remade itself into a major banking center greatly disproportionate to its size.

Urban hierarchies are never static but change constantly over time. Forces such as population shifts, immigration (including rural-to-urban migration), technological changes, and the evolution of the national and global economies cause some cities to grow, new ones to emerge, and older ones to decay. Capitalism is a dynamic, continually changing society, and geographically this results in new geographies of centrality and peripherality. In the 19th century, for example, Boston, Philadelphia, and New Orleans were major centers of international trade and immigration. Today their role has been largely eclipsed by the rise of newer centers in the American West, although Boston and Philadelphia are important in sectors such as health care, and New Orleans is still an important port. There is always movement up and down the hierarchy as cities gain or lose people and economic significance. Chicago in the mid-19th century was not particularly important; three decades later it had exploded to become a major manufacturing center and the second-largest city in the United States (Chapter 3).

5.2 Small Towns

At the bottom of the urban hierarchy lie small towns, which often receive less attention than large cities. But small towns are important, forming bridges between urban and rural life, offering a mixture of characteristics. Small towns often offer easy access to natural amenities that large cities do not.

People have lived in small villages since the Neolithic (Chapter 2), and many continue to do so today. In less urbanized societies, the majority of people continue to live in hamlets and villages.

The definition of a small town is slippery, and the defining criteria vary over time and among countries. Even within the United States the legal definition varies among states. Generally, the term "small town" is taken to mean an urban area of less than 10,000 people, but this is an approximation. Many started with an export base (Chapter 4) in agriculture. There is considerable variation in the size and type of small towns: some are thriving tourist centers; others are depressed farming communities. Some small cities, like Vail, Colorado, are extremely wealthy and expensive. Some are coastal fishing villages; others little more than a gas station serving passersby. Some have been swallowed by suburbanization on the urban periphery.

The texture of life in small towns is different from that in large cities. Social bonds tend to be more cohesive, there is a stronger sense of community, and more people know one another. The stereotype is that everyone knows everyone. People may, for example, know their neighbors well, and know other residents in several different roles (for example, one might become friends with their real estate agent). Small towns tend to be more ethnically homogeneous than cities. Trust flourishes in this context, though outsiders may be regarded with suspicion. Traditionally, small towns have been more religious and conservative than large cities, although not always. Public transportation is often nonexistent, although parking is rarely a problem. With less traffic, drivers spend less time in congested lanes, if any time at all. However, people living in small towns often need to drive to other urban areas for groceries or specialty stores, and some students must take buses to schools in other communities. Towns located in proximity to state or federal highways tend to do better economically than those located far away.

Small towns have often been idealized, especially in the United States, as offering an escape from urban ills, as places without crime and devoid of the problems that plague large cities. Small towns are often painted as affordable utopias where one can live a quiet life. This view often celebrates agricultural work and farming, which in reality is difficult and grueling. The romanticized idea has been popularized on countless television shows, and many Americans have a profound affection for small-town communities.

The reality is rather different. Crime rates do tend to be somewhat lower in small towns than in large cities, but many small towns suffer from property theft, alcoholism, gun violence, spousal abuse, and economic depression. Medical facilities may be difficult to access, especially given the closure of many rural hospitals. The lack of recreational opportunities can lead to boredom and frustration, although the Internet has changed this situation some. Agriculture has become so mechanized that it offers little employment. The lack of jobs drives many young people to leave, seeking their fortunes in larger cities as they move up the urban hierarchy. For young adults catapulted into large urban environments, the experience can be disorienting. The construction of the interstate highway system (Chapter 8) deprived many small towns of visitors and badly needed revenues. Many small towns, especially in isolated contexts, have been shrinking in population, and the remaining residents tend to be older adults.

5.3 Classical Approaches to Understanding Urban Hierarchies

Geographers have sought to comprehend the dynamics of urban hierarchies for a long time. These efforts gave way to three major approaches that are discussed here, including Central Place Theory, the rank-size rule, and the study of primate cities. Each is examined in turn.

5.3.1 Central Place Theory

An initial attempt to theorize urban hierarchies was Central Place Theory, developed by Walter Christaller as his doctoral dissertation *Central Places in Southern Germany* in 1933. It led to a highly influential set of models widely used to describe and analyze urban hierarchies in the mid-20th century.

Essentially, Central Place Theory is a model of city-systems that posits them as retail centers (central places) that distribute goods and services to their surrounding hinterlands. Like many models of spatial analysis, it assumes a featureless isotropic plain in which population density and transport costs are equal in all directions. Each good has a threshold, or minimum market size, as well as a range, or maximum distance consumers will travel to purchase it. The size of a given good's range and threshold vary with transportation costs, or the friction of distance, as well as the willingness of consumers to travel to obtain a given good or service. As the effective price increases with distance from the central place, demand declines accordingly. The model is thus compatible with neoclassical views of economics that privilege consumer demand in the analysis of social and spatial structures. Different goods and services have different thresholds: inexpensive, frequently purchased, everyday necessities have low thresholds (e.g., convenience stores, groceries, eating and drinking establishments, dry cleaning and other personal services), while costly, infrequently purchased goods and services, including luxuries, have much larger ones (e.g., jewelry, specialized medical services). The degree of specialization of a good or service is thus directly proportionate to the willingness of consumers to travel to obtain it: increasingly specialized products attract consumers from progressively longer distances. A hierarchy of goods and services leads to a hierarchy of central places to distribute them. Central Place Theory thus ranks urban places on their ability to distribute goods that range from "low order" to "high order" ones. The order of a central place is determined by the highest-order goods that it distributes to consumers around it.

The resulting system of central places—the urban hierarchy—reflects the imperatives of market forces over space, which tend to maximize the number of centers, maximize the number of consumers served, and minimize the aggregate distances that consumers must travel to purchase goods of varying levels of specialization. Large cities distribute both low and high-order goods, the latter over increasingly large hinterlands. Higher-order centers are more widely spaced, have larger populations, serve large market areas, and are encountered less frequently on the landscape. Thus, metropolitan areas, the highest order, offer the greatest diversity of goods and services and hinterlands that extend over vast distances. Conversely, progressively lower-order centers are nested within the market areas of higher-order centers. Small, rural hamlets, for example, historically formed central places serving low-density agricultural regions. Progressing up the urban hierarchy, one encountered increasingly larger but infrequent places with mounting diversity of goods and services.

In the Classical Central Place model, which has several variations, consumers minimize their transport costs to the closest city that offers the goods they desire, and producers maximize profits. On an isotropic plain, market areas form circles around each center. However, as consumers located between competing centers opt for the closest one, overlapping market areas are divided by lines. (Fig. 5.1). Assuming no spatial variations in transport costs or population, a hexagonal series of market areas should emerge, with lower-order ones nested within the hinterlands of higher-order ones. Hexagonal market areas became the defining feature of Central Place Theory, and were justified on grounds that they "packed" space, meaning that they serve all clients without overlapping market areas, more efficiently than any

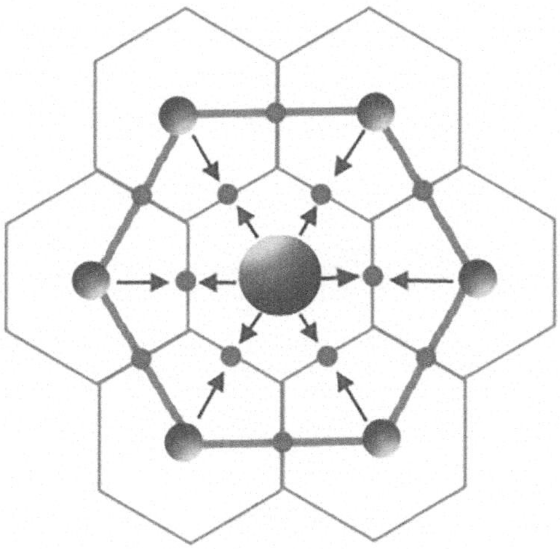

Figure 5.1
An example of how Central Place Theory holds that smaller centers emerge midway between larger ones as consumers minimize their travel costs.
(https://commons.wikimedia.org/wiki/File:Christaller_model_2.jpg.)

other shape. The model was thus decisively static in nature, and offered a fixed, frozen geometrical portrait of cities.

Central Place Theory became hugely popular in geography in the 1950s and 1960s as part of the emerging positivist school of thought, which emphasized modeling. It formed the dominant paradigm within urban geography for almost a generation. For example, urban geographers, notably Brian Berry, sought out hexagonal systems of market areas in the Midwest, which most closely resembled an isotropic plain. The model was also used in urban planning ventures in Israel and the Netherlands. Positivist theoreticians built increasingly more complex—and unrealistic—central place models that attempted to incorporate uncertainty, entropy, and evolution over time.

Ultimately, Central Place Theory was abandoned as human geography shifted decisively into political economy and social theory in the late 20th century. It was widely criticized for its overly simplistic assumptions; the absurdly unrealistic reduction of cities to shopping centers; the neglect of social relations in favor of individual consumer choice and resulting silence concerning class, gender, consciousness, and power; the absence of any account of production and the spatial division of labor; and its ahistorical portrayal of urban networks. In many respects Central Place Theory reflects much older urban systems with closed, local hinterlands that existed under feudalism. Today, the once-dominant model of urban geography remains little more than a historical curiosity.

5.3.2 Rank-Size Rule

Another attempt to grasp the dynamics of urban hierarchies is the rank-size rule, which states that the rank associated with some size *S* is proportional to *S* to some negative power. It was first proposed by Zipf in 1949. The rank-size rule is commonly expressed in the equation

$$R_i = \alpha S_i^{-\beta} \tag{5.1}$$

where R_i is the rank of the *i*-largest city, S_i is the city's size, and β is the exponent coefficient. A graphical expression of this relation is found in Fig. 4.10. When the negative power is close to one, the rule implies that the second-largest city is half the size of the first, and that the third-largest city is one-third of the size of the largest, and so on. In cases where this power is greater than one, it suggests that the second-largest city is more than half the size of the largest city, the third-largest city is more than one-third of the largest city, and so on (Fig. 5.2).

An example of the rank-size rule for Australia follows in Fig. 5.3. The largest city, Sydney, has roughly 4.3 million people. Theoretically, the second-largest city, Melbourne, should have one-half this population; Melbourne's actual population, 3.8 million, is thus larger than what

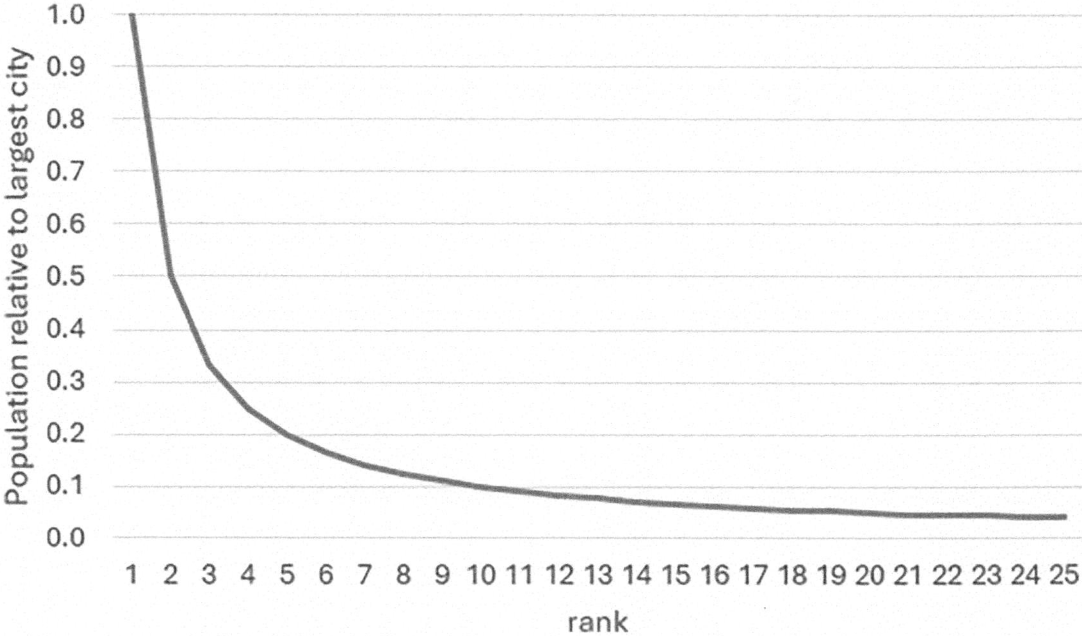

Figure 5.2
The rank-size rule.

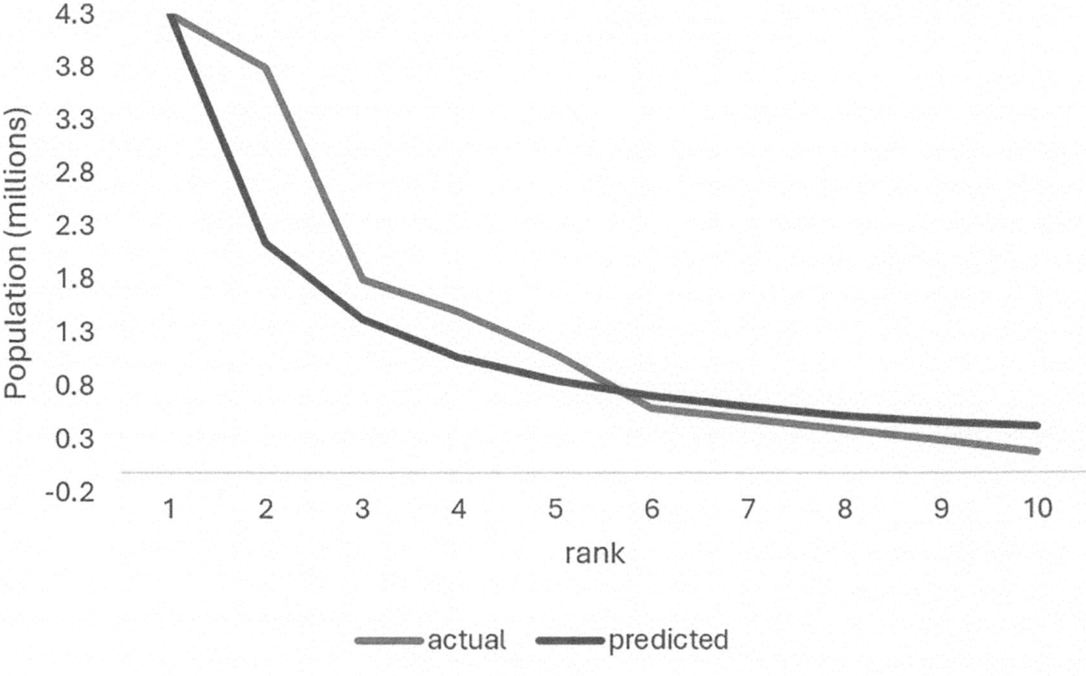

Figure 5.3
Empirical example of rank-size rule, Australia.

the rule predicts. Nonetheless, as one moves down the urban hierarchy, a rough equivalence between observed and predicted values emerges, indicating the rule has some empirical merit.

A log transformation can be used to estimate β as follows:

$$\log(R_i) = \alpha - \beta \log(S_i) \tag{5.2}$$

A logarithmic transform changes the curve into a line. Empirical estimates of the estimated coefficient range from 0.809 for Morocco to 1.963 for cities in Australia.

A striking feature of the rank-size rule is how well it fits across nations and over time. It has proven to be remarkably stable and consistent over many different contexts. Although the rule does not explain *why* this pattern holds, it nonetheless fits empirically in a wide range of contexts.

5.3.3 Primate Cities

A third attempt to understand urban hierarchies concerns primate cities. Primate cities are not only the largest centers within national urban systems, they are disproportionately large, far bigger than the second-largest (hence violating the rank-size rule). For example, Colombo, Sri Lanka, is 45 times larger than the next-largest city; Bangkok, Thailand, is 40 times larger than its next-largest rival. Table 5.1 offers several examples and their relative primacy (size relative to the second-largest center) from around the world.

Such cities may hold up to 40% or 50% of a country's population. They contribute enormously to national economies. Many are also national political capitals, offering ready access to government offices. Most, but not all, are found in developing countries. In the developed world, examples include Paris, France; Vienna, Austria; Stockholm, Sweden; Athens, Greece; and Berlin, Germany.

The search to explain the rise of primate cities focuses on the enormous agglomeration economies they offer, that is, their capacity to reduce costs and improve productivity through proximity. Once started, they initiate self-reinforcing cycles of growth, drawing capital and migrants from surrounding areas. They often have an infrastructure superior to that of competing cities.

5.4 Interurban Flows

Because cities never exist in isolation of one another or their hinterlands, the flows of capital, people, goods, and information among them are critically important. Flows among cities reflect, and in turn shape, the uneven distribution of different economic sectors, jobs, and opportunities. A city like New York, for example, could hardly feed itself, but does well

Table 5.1 Examples of primate cities.

Country	Primate City	Primacy
Algeria	Algiers	5.1
Argentina	Buenos Aires	8.3
Austria	Vienna	8.6
Bangladesh	Dhaka	4.3
Chile	Santiago	6.5
Colombia	Bogotá	3.0
Cuba	Havana	4.9
Denmark	Copenhagen	6.1
Ethiopia	Addis Ababa	9.8
Finland	Helsinki	3.9
France	Paris	5.5
Greece	Athens	3.5
Iceland	Reykjavik	11.5
Iran	Tehran	4.3
Iraq	Baghdad	4.5
Jamaica	Kingston	3.2
Jordan	Amman	5.9
Kenya	Nairobi	3.9
Lebanon	Beirut	7.6
Liberia	Monrovia	26.8
Mexico	Mexico City	4.1
Mongolia	Ulaanbaatar	15.1
Nepal	Kathmandu	9.8
Nicaragua	Managua	12.4
Panama	Panama City	7.4
Philippines	Manila	4.3
Rwanda	Kigali	12.6
Sri Lanka	Colombo	45.0
Tanzania	Dar es Salaam	4.9
Thailand	Bangkok	13.4

economically as a center of banking, reliant on its connections to other parts of the United States and the global economy for its necessities. Several technologies have greatly facilitated the explosion of interurban flows over the years.

The completion of the transcontinental railroads in the 1860s had enormous impacts on the American urban hierarchy, accelerating flows of people, goods, and information among cities (Chapter 3). Railroads led to the growth of new cities in the West, the explosion of Midwestern manufacturing centers, and provided supplies and raw materials to reach the East Coast. Today rail lines continue to be vitally important. Because ports are the major points of entry for imports, rail and trucking routes from ports to other cities are particularly important. These forms of transport are relatively cheap, which helps to hold down the prices of consumer goods.

Along with the automobile, trucks became widespread in cities during the 1920s. These greatly facilitated the movement of goods among cities, allowing the spatial division of labor to unfold more rapidly. The construction of the US Interstate Highway System had major impacts on the American city-system (Chapter 3 and Chapter 8). Among other things, it allowed trucks to move quickly and relatively safely among cities, deepening their connections with one another. By reducing transport costs, trucks facilitated the rise of urban comparative advantages.

Flows of goods were always accompanied by flows of information. In the mid and late 19th century, the telegraph became vital to the growth of firms with plants and offices in more than one city. The telegraph allowed vast movements of information about crop prices, jobs, and financial investments, reducing uncertainty and improving productivity. Similarly, at the end of the 19th century, the telephone started to come into widespread use, although few households could afford one in their homes until the 1920s.

After World War II, air travel started to become cheaper and more affordable for more people. Given the rising value of time that accompanies higher incomes, more people could choose to fly between cities rather than drive. The airline system has thus become a vital source of interurban movements of people (Fig. 5.4) as well as high-valued goods (low-valued goods move by truck, ship, or rail). Interurban patterns of flights reflect the respective sizes of cities, although the widespread use of the hub-and-spoke system by different airlines complicates that pattern.

The movements of people and goods among cities exhibit a surprisingly regular pattern that has been modeled after gravity. In the 17th century, the British genius Isaac Newton theorized the law of gravity, holding that the attraction between two bodies is proportionate to their mass and inversely proportionate to the square of the distance between them. Similarly, movements among cities tend to be proportionate to their respective populations and the distance between them. Mathematically, the gravity model may be expressed as:

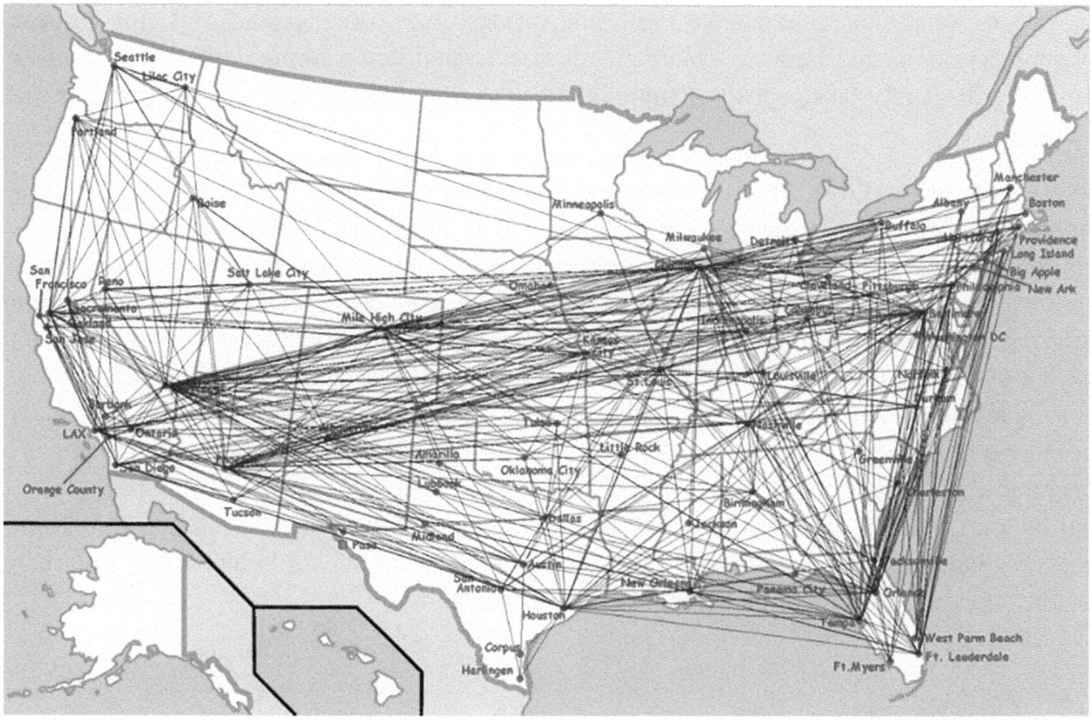

Figure 5.4

Route map for Southwest Airlines.

(https://commons.wikimedia.org/wiki/File:Southwest_Airlines_Route_20110327.png.)

$$M_{ij} = (P_{i*}P_j)/d^k,$$

where M_{ij} is the number of people moving between cities i and j, P is their respective populations, d is the distance between them, and k is an exponent. (In the original gravity model k was equal to 2, so gravity declined by the square of distance, but in population models the exponent has more possibilities). This means that there will be more movements and interactions between large cities than small ones; moreover, interactions will decline between places as the distance that separates them increases, reflecting the friction of distance. This pattern has been observed and confirmed in a wide variety of empirical contexts. The gravity model is widely used today by traffic planners to understand and predict flows among places, including patterns of migration and retail trade.

It is important to note that as new infrastructures and technologies have reduced transport times and costs, the friction of distance has declined. Transport costs were once a major cost to firms selling goods; today they are a small share of the final price. Trucking, shipping innovations (e.g., containerization), rail innovations (e.g., double stacking), just-in-time inventory systems, and highways have greatly accelerated flows among cities. One result is that many firms rely on fewer local suppliers, resulting in a "hollowing out" of many cities' economies.

Finally, interurban flows today include the prodigious amounts of information that flow through fiber optic cables. Starting in the 1980s, many cities began to lay fiber in vast amounts to connect people and firms within cities and also among them. Because contemporary capitalism is so information-intensive, urban growth has been greatly accelerated by a worldwide digital telecommunications network, notably the fiber optic system that forms the nervous system of the world economy. This technology enabled the Internet as well as global cable television and international news services; news travels around the world at unprecedented speeds. However, the uneven spread of fiber, the Internet, and cell phones generates their own geographies. Access to communications is restricted by the uneven distribution of wealth all over the world. Even within cities, there are uneven geographies of access, including offline places and peoples, and there are profound differences between urban and rural places in terms of their digital connectivities.

5.5 City-Regions

The geography of cities reflects two sets of forces that profoundly shape local patterns of employment, housing, and land use. On the one hand are agglomeration economies (Chapter 4), which lead to the centralization of many functions, particularly those that rely on dense interactions with other firms and face-to-face contact. These interactions are the lifeblood of the city. On the other hand, there are forces that propel decentralization, including expanding commuter sheds and the growth of telework, which have undercut agglomerative advantages in some cases. Relentless exurbanization, led by housing, is a distinctive sign of the latter. City-regions reflect this tension, noting that the dispersal of growth to the metropolitan periphery has greatly enlarged the size of many urban areas. City-regions represent a strategy under capitalism to maximize opportunities for local interactions and yet avoid the high costs that typically accompany dense concentrations of activity in downtown areas. The rising importance of city-regions reflects what economists call network externalities, or the benefits that accrue from being locked into mutually beneficial relations of interdependence.

City-regions are networks of cities in a given region with close functional ties to one another. These urban formations come in a wide range of sizes, but typically contain within them several million people. Typically a city-region is dominated by a large metropolitan area, with smaller cities and towns on the periphery. The dominant city acts as an anchor that stabilizes regional growth, and its CBD serves as a gravitational core. The economic base (Chapter 4) of the largest city links the region to global circuits of capital, and it may be the gateway through which immigrants pass. Subcontracts by firms in the metropolitan core fuel corporate growth in other parts of the surrounding area and economy.

As we have seen, cities are always linked to their surrounding hinterlands. While Central Place Theory sought to describe this process purely in terms of consumption, more recent approaches emphasize a much broader array of interactions. Cities develop functional hinterlands that include commuting areas, shopping patterns, local purchases of business services, housing, catchment areas for airports, social networking, and markets for local produce and goods. These broader places are called city-regions (at times they may be called mega-urban regions or polycentric metropolises). The idea is not new, and may be traced back to a 1915 book by Patrick Geddes called *Cities in Evolution*. After World War II these regions grew rapidly in size, number, and importance. As growth has spilled far outside the boundaries of central cities, the idea of city-regions has become increasingly popular.

For example, England, the most populous part of the United Kingdom, exhibits several major city-regions, the largest of which is London (Fig. 5.5). City-regions may grow and merge into one another, giving rise to a megalopolis (Chapter 3).

In the United States, city-regions are delimited for statistical purposes as Standard Metropolitan Statistical Areas (SMSAs), which are defined as areas that contain a city of at least 50,000 people and surrounding communities with a high degree of social and economic dependency (a definition designed to capture commuting sheds). SMSAs contain the vast bulk of the American population (Fig. 5.6).

City-regions are heterogeneous, reflecting the deep imprints of their historical trajectories and varying roles in national and world economies. There are profound differences among them in terms of their local economies, amount and nature of foreign investment, immigration, and trade relations. Moreover, city-regions are not simply passive recipients of global processes but exert agency in their own right, as reflected in diverse local land use and infrastructure policies. Some city-regions are highly globalized. Some act as sanctuary cities for undocumented migrants. Some have an economic base in high-technology industries, while others are older, struggling deindustrialized centers. Most are important nodes in increasingly globalized value chains. Thus, while city-regions all exhibit dense local networks among their various urban areas, there is no one-size-fits-all model that encapsulates their variations.

City-regions pose problems for regional governance. A region may consist of numerous politically independent municipalities that form a mosaic or archipelago of places that simultaneously cooperate and compete with one another. Yet the problems they face affect all of them unequally. Local growth coalitions are insufficient to solve region-wide problems such as congestion and high housing costs. Encouraging coordination among cities in a region is often difficult, especially when they have varying priorities and political agendas. City-regions are important to urban planners as they reflect how urban social relations are not

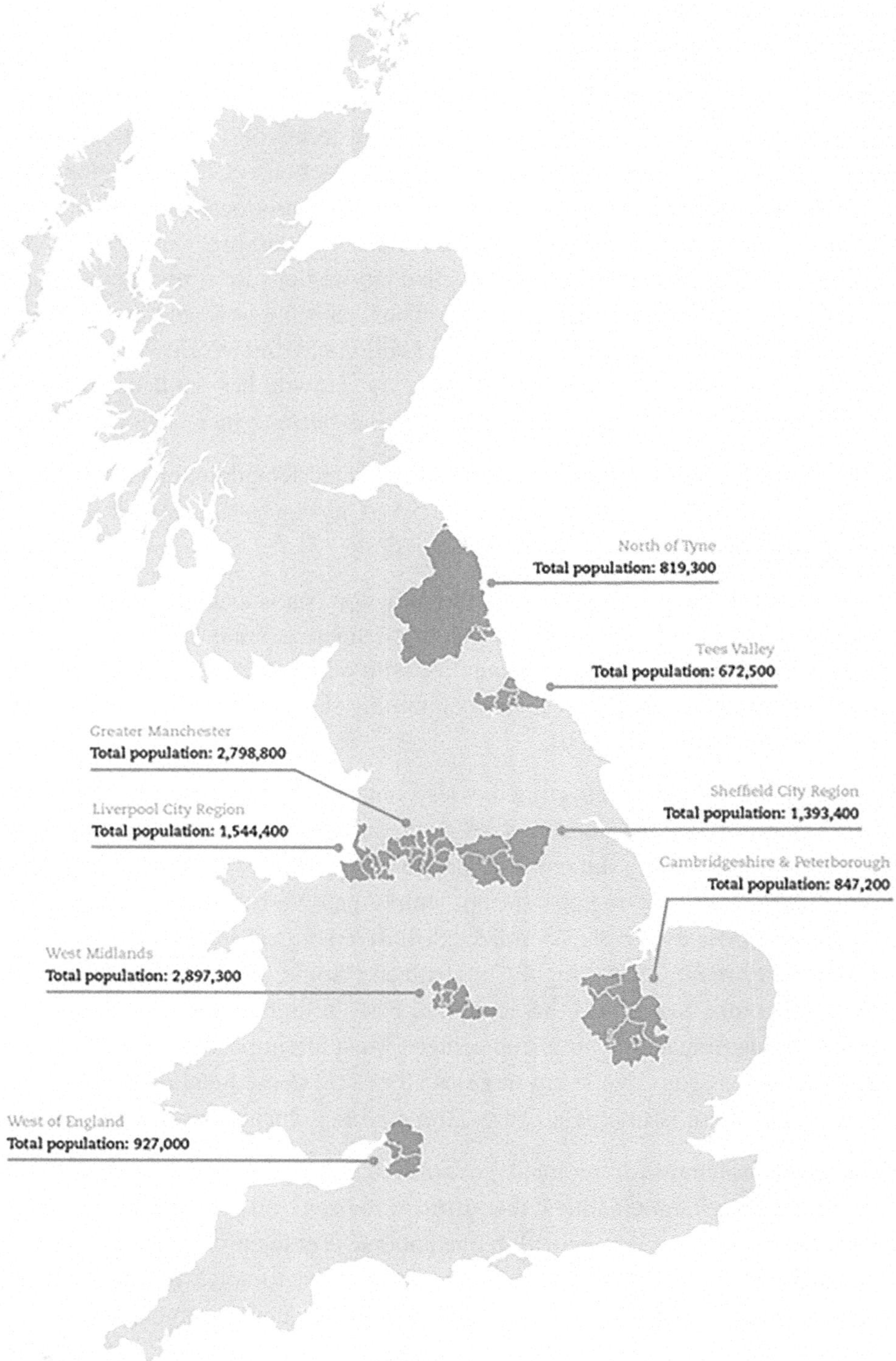

Figure 5.5
City-regions of England. Map lines delineate study areas and do not necessarily depict accepted national boundaries.
(https://commons.wikimedia.org/wiki/File:2019-Metro-Mayor-Regions-Map-01.png.)

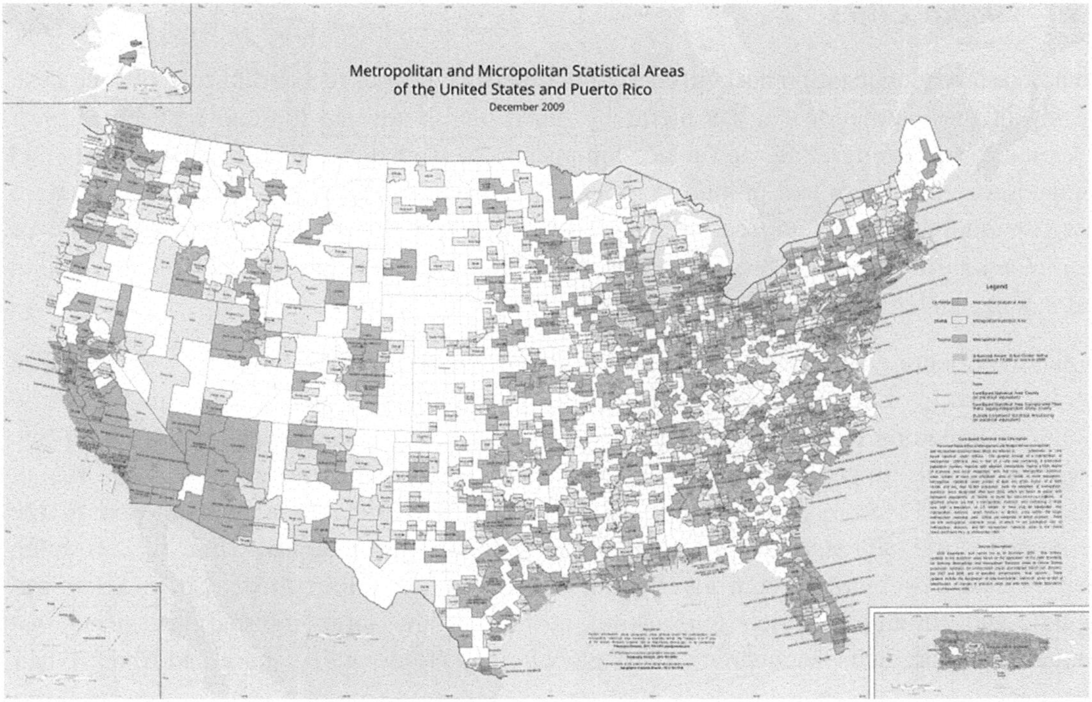

Figure 5.6
US standard metropolitan statistical areas. Map lines delineate study areas and do not necessarily depict accepted national boundaries.
(https://commons.wikimedia.org/wiki/File:Metropolitan_and_Micropolitan_Statistical_Areas_of_the_United_States_and_Puerto_Rico.svg.)

confined to one place. Air and water pollution, for example, typically manifest in broad regions that include numerous cities. The widespread nature of these problems invites planners to think about public investments, infrastructure, and public spaces (e.g., green belts) on a regional rather than local level. In a very real sense, "urban" issues and problems have become decisively regional in scope.

City-regions have proliferated around the world. Today there are thousands of city-regions scattered across the globe. In rapidly developing economies such as China, city-regions include the metropolitan areas of Beijing and Shanghai, as well as the vibrant Pearl River Delta centered on Hong Kong.

In many respects city-regions have become the building blocks of the world economy. Their growing significance has led to speculation that the world economy today is more centered on a necklace of globalized city-regions than the nation-state, which has dominated global geopolitics since the 17th century.

5.6 Global Cities

Contemporary globalization has witnessed the growth of a series of interlinked global cities at the top of the international urban hierarchy. Such places are tied through vast tentacles of investment, trade, migration, and telecommunications to clients and markets, suppliers and competitors, consumers and producers around the world. Global cities serve as the home to massive complexes of financial firms, producer services, and corporate headquarters, or "command-and-control" centers in the world system. Through their vast webs of information and capital, global cities drive the world economy.

Typically, London, New York, Paris, and Tokyo are positioned at the top of the world's urban hierarchy, with cascading layers demarcated by successively smaller roles in the world economy, including Frankfurt, Toronto, Los Angeles, Chicago, San Francisco, Osaka, Hong Kong, Shanghai, Singapore, and Abu Dhabi. Others have pointed to cities such as Miami, Madrid, Milan, Moscow, and Beijing. Global cities are primarily a phenomenon of the developed world, although they are appearing in several developing ones (Fig. 5.7). Through their extensive webs of foreign investment, flows of labor, information, commodities, and capital, and through the decision-making powers of the large transnational firms that frequently locate their headquarters there, global cities "fold" global space and time to their advantage.

The cores of such conglomerations allow for dense networks of interaction necessary to the performance of headquarters functions, including: monitoring frequent changes in niche

Figure 5.7
Map of global cities.
(https://commons.wikimedia.org/wiki/File:Global_Cities_Index.png.)

product markets; negotiating with labor unions; keeping abreast of new technologies and government regulations; keeping an eye on the competition; staying tuned to a complex financial environment; initiating or resisting leveraged buyouts and hostile takeovers; seeking new investment opportunities and so forth. Global cities serve as hubs of corporate headquarters and the vast webs of information that link American metropolitan regions together.

The crux of global cities' role in the contemporary world economy is to serve as arenas of interaction, allowing for face-to-face contact, political connections, artistic and cultural activities, and elites to rub shoulders easily and exchange tacit knowledge. At their core, global cities allow the generation of specialized expertise upon which so much of the world economy depends. However, there are important class dimensions to this process: if professionals and elites rely upon face-to-face contact with one another, the day-to-day conditions of the service class are often quite different, and their interethnic interactions may be relatively limited or even occasionally hostile in nature.

Much of the impetus to agglomerate in global cities, despite their manifestly high rents and labor costs, rests on the advantages procured through frequent face-to-face contact, which remains the most efficient and effective means of obtaining and conveying irregular forms of information, particularly when it is highly sensitive in nature. The elites of global cities rely heavily on interpersonal contacts saturated with trust and reciprocity to "get things done." Many such interactions are international in scope, necessitating an appreciation of foreign cultures and languages, as does managing a multiethnic labor force. Assets such as thick networks of interpersonal ties, with all of the issues of trust and external economies of scale that they embody, are very difficult to reproduce in other contexts, an observation that goes far to explain why the primacy of global cities such as Amsterdam, London, and New York has changed very slowly over the last 500 years. The growth of the global telecommunications network has not, contrary to early popular expectations, undermined the status of global cities, but in fact enhanced their stature, particularly when the information produced and exchanged is sensitive, irregular, proprietary, and unstandardized in nature. The Internet and global fiber optics may have increased rather than decreased the need for face-to-face contact: as oceans of ceaselessly circulating data wash over global cities, the interpretive ability to extract meaning from them becomes ever more pressing. For this reason, most managers spend the bulk of their working time engaged in meetings in which the exchange of "buzz," or tacit knowledge, is critical. Indeed, financial and business services firms not only pay high rents to be near city centers, and endure the severe congestion such locations often entail, but spend lavishly to fly their executives around the world to meet with their counterparts in person.

Given that globalization has reached ever farther down urban hierarchies across the globe, to some extent *all* cities are global cities in that they are inescapably enveloped in worldwide networks of goods, people, capital, and information. Indeed, many smaller cities and even rural areas are also undergoing profound transformations as a result of the information technology revolution.

Global cities are, however, much more than simply centers of capital accumulation. They are also important centers of discourse and the production of culture. Because their *raison d'etre* cannot be reduced to the "economic," but includes a vast variety of formal and informal cultural and political interactions such as tourism, publishing, the printed and electronic media, film and cinema, education, and fashion, global cities are much more than simply poles for the production of corporate knowledge. Precisely because they exert so much influence over their respective national economies, and the global one, global cities are also motors of ideological change, absorbing waves of immigrants, socializing people in complex ways around the crisscrossing currents of cultural difference, and generating political agendas that reverberate to shape the possibilities and potentials of action across the globe.

5.7 Chapter Summary

Cities are embedded within networks of flows, movements, and interactions of people, goods, capital, and information that suture them together. Systems of cities of various sizes are called urban hierarchies, which include settlements ranging from small towns and villages to enormous urban centers.

Traditional approaches to the study of urban hierarchies began with the hugely influential, but now outdated, Central Place Theory, invented by Walter Christaller. Central Place Theory portrayed cities in terms of consumption and the market areas around cities, but ignored social relations, historical context, production systems, and changes over time. The rank-size rule mirrors statistically the size relations of cities in relation to the largest urban area within a given system, a pattern that has held up remarkably well in many different contexts. The primate city notion examines urban hierarchies dominated by a single, enormous urban area that outshines all others in size (e.g., Mexico City, Havana, Bangkok, Seoul, or Paris).

Interurban flows, including people and trucks moving on highways, airline flights, and streams of goods, money, and information, allow cities to specialize around their economic or export base (Chapter 4). The gravity model, which holds that the size of flows between two cities is proportionate to their size and inversely related to the distance between them, has been widely used to analyze this issue.

City-regions are areas in which local urban areas interact densely with one another. In the United States, this notion is captured through Standard Metropolitan Statistical Areas (SMSAs), regions with a city of at least 50,000 people and surrounding places that are functionally connected.

Finally, as urban hierarchies have become steadily globalized, they have been steadily dominated by global cities, large, dynamic, and influential centers that act as the command-and-control centers of the world economy. Global cities are as old as capitalism, including, for example, Amsterdam and London. However, the number, size and importance of global cities has grown rapidly over the last century. The significance of global cities lies not in their population size—there are many large cities in the global south that are not global cities—but in their economic significance. Global cities are centers of international banking, finance, and investment, with large stock markets. They are the places where many transnational firms locate their headquarters. They are usually important in shaping the global media (publishing, movies) and are centers of fashion, culture, and tourism. Some maintain that the world's networks of globalized city-regions herald the geography of the future, challenging the nation-state as the preeminent geopolitical entity on Earth.

Study Questions

1. What is an urban hierarchy?
2. What was Central Place Theory?
3. What is the rank-size rule?
4. What is a primate city? Give three examples.
5. What is the gravity model of migration?
6. What is a city-region? Give an example.
7. What are global cities and what role do they play in the world economy?

Further Reading

L. Andres, J.R. Bryson, W. Graves, B. Warf, Urban value chains and re-framing agglomeration-centric conceptions of urban theory. Routledge, United Kingdom Urban Geography. 02723638. 44 (8) (2023), 1701–1722, Available from: http://www.tandfonline.com/loi/rurb20?open=34&repitition=0#vol_34, https://doi.org/10.1080/02723638.2022.2125665.

Jonathan V. Beaverstock, Richard G. Smith, Peter J. Taylor, World-City Network: A new metageography?. Annals of the Association of American Geographers. 0004–5608. 90 (1) (2000), 123–134, Available from: https://doi.org/10.1111/0004–5608.00188.

John R. Bryson, Ronald V. Kalafsky, Vida Vanchan, Ordinary Cities, Extraordinary Geographies: People, Place and Space. Edward Elgar Publishing, Edward Elgar Publishing, (2021), Available from: https://doi.org/10.4337/9781789908022.

N.M. Coe, P. Dicken, M. Hess, H.W.C. Yeung, Making connections: Global production networks and world city networks. Global Networks. 14710374. 10 (1) (2010), 138–149, Available from: https://doi.org/10.1111/j.1471–0374.2010.00278.x.

B. Derudder, P. Taylor, P. Ni, A. de Vos, M. Hoyler, H. Hanssens, D. Bassens, J. Huang, F. Witlox, W. Shen, X. Yang, Pathways of change: Shifting connectivities in the world city network, 2000–08. Urban Studies. 1360063X. 47 (9) (2010), 1861–1877, Available from: https://doi.org/10.1177/0042098010372682.

J. Friedmann, G. Wolff, World city formation: an agenda for research and action. International Journal of Urban and Regional Research. 14682427. 6 (3) (1982), 309–344, Available from: https://doi.org/10.1111/j.1468–2427.1982.tb00384.x.

A. King, Global Cities. Routledge, Routledge, (2015).

Zachary P. Neal, From central places to network bases: A transition in the U.S. Urban Hierarchy, 1900–2000. City & Community. 1535–6841. 10 (1) (2011), 49–75, Available from: https://doi.org/10.1111/j.1540-6040.2010.01340.x.

John Parr, Perspectives on the city-region. Regional Studies. 0034–3404. 39 (5) (2005), 555–566, Available from: https://doi.org/10.1080/00343400500151798.

Saskia Sassen, The Global City: New York, London, Tokyo. Princeton University Press, Princeton University Press, (2001), Available from: https://doi.org/10.1515/9781400847488.

Allen J. Scott, City-regions reconsidered. Environment and Planning A: Economy and Space. 0308–518X. 51 (3) (2019), 554–580, Available from: https://doi.org/10.1177/0308518x19831591.

A. Scott, Global City-Regions: Trends, Theory and Policy. Oxford University Press, Oxford University Press, (2002).

D. Smith, The world urban hierarchy: Implications for cities, top to bottom. Brown Journal of World Affairs. 11 (2) (2005), 45–55.

P. Taylor, B. Derudder, World City Network: A Global Urban Analysis. London: Routledge. London: Routledge (2016).

N. Zdanowska, C. Rozenblat, D. Pumain, Evolution of urban hierarchies under globalization in Western and Eastern Europe. Hungarian Central Statistical Office, United Kingdom Regional Statistics. 20648243. 10 (2) (2020), 3–26, Available from: http://www.ksh.hu/terstat_eng, https://doi.org/10.15196/RS100202.

CHAPTER 6

Urban Residential Space

Chapter Outline

Cities are not only production machines, they are places where people live and experience everyday life, construct their biographies, raise families, and embody their hopes, aspirations, and fears. Put differently, in addition to the systems of production that center upon profit-maximization, cities are also networks of social reproduction. It should be emphasized that residential trends never occur without linkages to the urban economy; put differently, residential space and urban production space are always intertwined.

This chapter examines urban residential space in several steps. It begins with a review of a very famous and influential body of work, the Chicago School, which essentially started urban geography. Second it turns to housing markets, the systems of supply and demand that produce housing of different types at different prices in different parts of the city. Third, it examines neighborhoods, which are important institutions structuring everyday life in the city. The fourth part focuses on suburbs, the dominant trend of urban growth in many countries. Fifth, it discusses gentrification, the growth of relatively affluent areas near the city center that has been driven by the surge in producer services over the last few decades. The sixth part addresses the urban ghetto, a term originally used for Jewish quarters in European cities but that in the United States has come to mean parts of the inner city dominated by low-income minorities. The last section summarizes homelessness, or people excluded from housing altogether.

Urban Geography. DOI: https://doi.org/10.1016/B978-0-443-30281-7.00006-2

6.1 Social Ecology and the Chicago School

Urban geography as an academic discipline and body of research largely began in Chicago during the 1920s. As the model of the rapidly growing, industrialized city populated by streams of immigrants, Chicago became the prototypical example of the industrial American city. The University of Chicago played a major role in disciplines such as economics, sociology, and geography. Within this context, the Chicago School of urban studies arose, which was enormously influential in sociology and geography for the next several decades. The Chicago School is credited with the first systematic attempt to understand the dynamics of urban areas, including social change, urban planning, and territoriality.

The origins of the Chicago School lay largely with Robert E. Park, a former journalist turned teacher. In 1925, Park, E. W. Burgess, and Roderick McKenzie published *The City*, a famous collection of interpretive essays about urban life, a volume that both summarized and inspired a long tradition of urban ethnography. Chicago School practitioners inaugurated the tradition of detailed case studies, ranged far and wide over the city, studying, among other things, the wealthy, immigrants, hoboes, the destitute, dance halls, criminals, and prostitutes in an attempt to draw rich and detailed portraits of urban life. In the process, they irrevocably fused the study of space and the study of society.

The first paradigm of urban structure offered by Chicago School theorists, particularly McKenzie, centered on the metaphor of the city as urban jungle, a view derived from the social Darwinism prevalent in the early 20th century. For example, the displacement of one ethnic group by another in a given neighborhood was framed as a process of invasion and succession, a view that drew directly from ecological studies of how one plant species displaced another through successive stages in the evolution of ecosystems. Later, this biological metaphor was dropped in the face of criticisms that it lacked a coherent account of social relations and naturalized the inequality of urban areas. Throughout the Chicago School's worldview, competition appears repeatedly as a driving force behind ethnic and class segregation.

Chicago School theorists also drew on the urban sociology of Ferdinand Tönnies and notions such as *Gemeinschaft* and *Gesellschaft* to examine daily life in light of the massive rural-to-urban migration then characteristic of most US cities. In this view, urbanization represented the annihilation of mythologized rural communities in which everyone knew everyone else. In contrast to the idealized image of small towns in which everyone ostensibly was intimately connected to everyone else, and presented the same sense of self under all contexts, urbanization was held to decompose these traditional bonds and erode the foundations of mutual trust. Cities, it was held, were not conducive to the formation of a sense of community. Louis Wirth, in particular, advocated a desolate but compelling view of city life as structured around population size, density, and heterogeneity. Size or total population, he held, created a climate that was inherently predatory, utilitarian, uncaring, and commodified: strangers were rare in small towns, but the norm in large cities. Density, he argued, led people to be close physically but not emotionally; indeed, alienation was the norm. Finally, social and cultural heterogeneity, manifested in the diverse lifestyles found in large cities, undermined the common values necessary to the success of healthy communities. Wirth concluded that these factors were responsible for the high rates of urban crime and other social pathologies ranging from suicide to psychoses. (Subsequent work has rectified this stereotype by pointing to the high crime rates often encountered in many small cities and the presence of healthy, vibrant urban neighborhoods.)

The most famous product of the Chicago School are three models of urban social structure endlessly repeated in introductory sociology and geography textbooks. The first, proposed by Earnest Burgess in 1927, was the Concentric Ring model (Fig. 6.1), which, extrapolating from the specific instance of Chicago, viewed the city as a series of rings of varying size centered on the Central Business District (CBD, a term coined by the Chicago School). Adjacent to the CBD was a zone of factories and warehouses, sometimes called a "zone of transition." Moving outwards, this was followed by the "zone of workingmen's homes," meaning working-class, blue-collar communities. Yet farther out were the medium, then high-income belts of suburbia. Burgess observed that cities tend to expand horizontally as the wealthy had new homes constructed on the urban periphery; as they came to occupy these, the relocation of families outwards set off a chain of vacancies that reverberated across the urban landscape, as less well-off families in turn occupied the cast-off mansions of the rich, a process known as filtering or the trickle-down theory of housing supply. Moreover, Burgess observed a paradox, that low-income residents in cities lived on expensive, accessible land near the urban core whereas more well-to-do inhabitants of wealthier rings occupied less expensive land. This puzzle, he noted, was explained by the population density curves characteristic of cities, which decline exponentially with distance from the CBD. The poor, crowded into dense communities in the urban core, collectively generate high aggregate rents that create relatively high rates of profit in inner city areas, whereas the low-density environments of the wealthier classes reduce the profitability of the less accessible periphery.

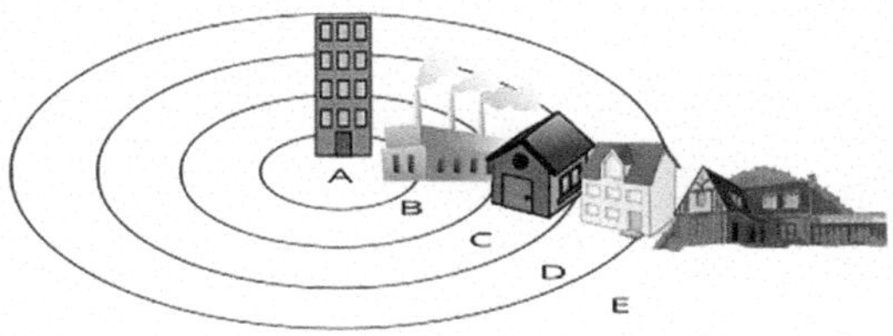

Figure 6.1
Burgess's concentric ring model.
(https://commons.wikimedia.org/wiki/File:Burgess_model.svg.)

In contrast to the rigid geometry of the Burgess model, Homer Hoyt, an economist and another influential Chicago School theorist, proposed the sector model of urban growth in 1939 based on an empirical analysis of 142 cities (Fig. 6.2). In this view, rather than concentric rings, urban growth occurred along transportation lines centered on the CBD. Once parts of the central city acquired distinctive uses, they radiated outwards. High-income land uses played a determining role in shaping the rest of the city, growing along waterfronts, high-altitude areas, or toward other high-income neighborhoods, and other uses filled in the spaces between them. Rather than belts, socioeconomic groups occupy sectors or wedges. The wealthy, with the greatest ability to pay, out-competed less advantaged groups for the most desirable locales, generally far from the disamenities of factories or railroad lines. Low-income groups found themselves confined in zones relatively far from the wealthy.

The third model of the Chicago School trilogy, proposed by Chauncy Harris and Edward Ullman in 1945, was called the Multiple Nuclei Model (Fig. 6.3). Essentially, this view attempted to rectify the perceived simplistic shortcomings of the previous two. Rather than a single city center, it maintained that American cities had become polycentric, with many nuclei, around which land uses were organized in a complex quilt. Rather than a single overarching logic, this perspective maintained that certain land uses would repel one another, while others might be mutually attractive.

Although the Chicago School began to diminish in importance shortly after World War II, its ideas were carried into urban sociology and geography for many years afterwards. For example, social ecologists in the 1960s, armed with multivariate statistical methods and census data, argued that each of the three classic models effectively captured a different aspect of urban social space. Thus, family status was distributed in rings per Burgess's Concentric Ring model, reflecting the dynamics of the family life cycle. Economic class was held to occur in sectors conforming to Hoyt's theory of land use. Finally, ethnicity was theorized to

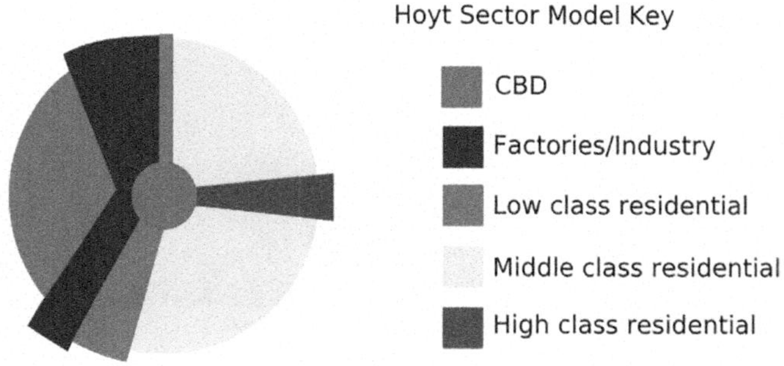

Figure 6.2
The sector model.
(https://commons.wikimedia.org/wiki/File:Hoyt_model.svg.)

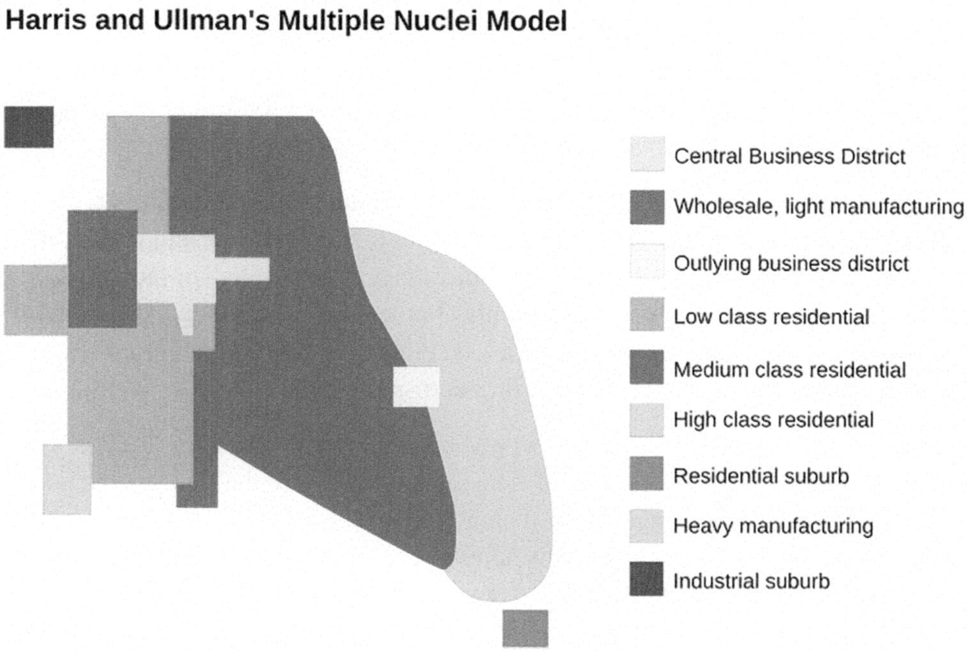

Figure 6.3
The multiple nuclei model.
(https://commons.wikimedia.org/wiki/File:Multiple_nuclei_model.svg.)

reflect the dense nucleations of different immigrant groups, as proposed by the Multiple Nuclei Model.

This approach has been heavily criticized and now is seen as obsolete. Pioneering though it was in the early days of urban analysis, Chicago School social ecology was simplistic. It offered a weak conceptualization of social relations, notably class and gender, and underplayed the political dynamics at work in urban change.

The Chicago School essentially defined American urban analysis throughout the 20th century. More recently, however, attempts by urban political economists to reveal the complex dynamics of globalization and immigration in a postindustrial context have yielded the so-called Los Angeles School, which takes the low-density, polynucleated, postindustrial southern California metropolis as its point of departure.

6.2 Housing Markets

In a capitalist economy, housing is a commodity, the price and availability of which are determined by supply and demand in housing markets. The state also provides housing, such as public housing projects (Chapter 8). Although housing is a commodity, it differs from other goods in that it is a necessity, nonmovable or fixed in location (only 2% of the United States lives in mobile homes, and most are not mobile), expensive (the greatest asset for most families), and very durable. Housing is also more than a commodity: it reflects its inhabitants' status, emotional identity, and social aspirations.

Housing markets are deeply shaped by national and even international trends, such as interest rates, inflation, and immigration. However, there is not one unified housing market but many local markets. The supply and demand for housing are segmented based on differences between ownership and rentals; single-family homes and apartment or condominium buildings; and their locations in central cities, suburbs, or rural areas. For these reasons, there are large variations in the cost of housing across the country.

There are numerous actors involved in the supply and demand for housing. The supply reflects both new construction and the filtering of housing (as noted by the Chicago School social ecologists). Landlords thus range from being single proprietors to large corporations. Conversely, the demand for housing reflects population growth in an area, its age composition (e.g., older people have different needs than young families with children), and local incomes. The demand for housing reflects buyers' subjective utility of units, which in turn include both the characteristics of the structure (its age, number of bedrooms, bathrooms, etc.) and its local neighborhood (quality of schools, proximity to commercial areas).

The median price of housing has risen steadily over time (Fig. 6.4). The steady increase reflects changing population levels and the fact that there has been insufficient construction of new homes to keep up with demand. Price increases have been particularly notable in large metropolitan areas, notably in California and the northeastern megalopolis.

In the United States, a wealthy country, most people live in housing that they own (Fig. 6.5). Typically, people with lower incomes rent their units and middle class and wealthier people own. Home ownership has long been a sign that residents had entered the middle class. To

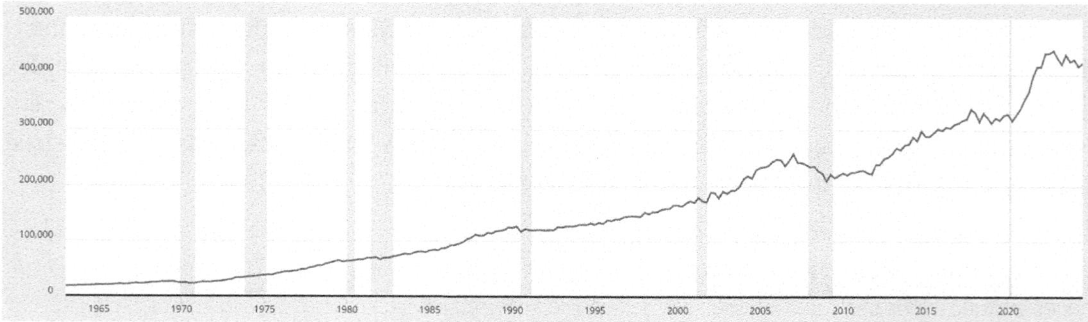

Figure 6.4
Median cost of housing in the United States over time.
(https://fred.stlouisfed.org/series/MSPUS.)

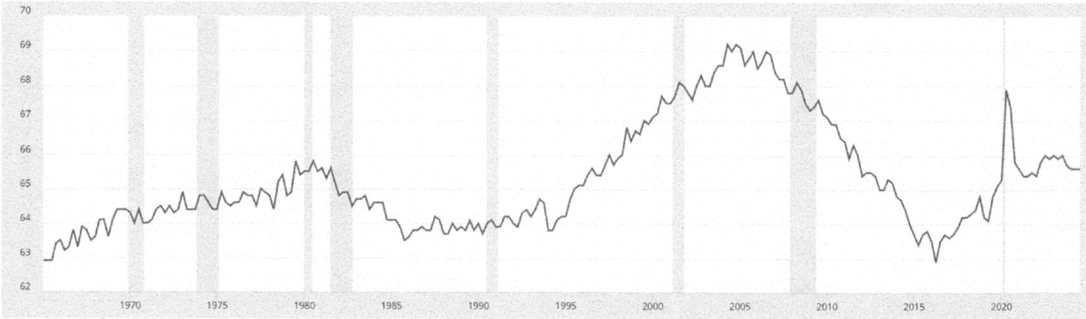

Figure 6.5
Share of the United States living in owned housing.
(https://fred.stlouisfed.org/series/RHORUSQ156N.)

encourage ownership, the federal government in the 1930s began to allow home owners to deduct the interest on mortgages from their taxable income, a subsidy to the middle class that is unique to the United States. However, ownership rates fluctuate over time and across the country, with wealthier areas having higher rates. In many large, expensive metropolitan areas where relatively few people can own homes, renters are the majority of the population and a potent political force.

Because housing is expensive, housing markets are closely intertwined with financial ones. Most people must buy a home by obtaining a mortgage from a lending agency, such as a bank or credit union. Lending agencies judge applications for mortgages based on the prospective buyers' incomes and credit worthiness. However, there is a long history of racial discrimination in this regard: traditionally many banks would not extend credit to ethnic minorities, particularly African Americans, in a process known as redlining. Although technically illegal today, redlining persists informally, making it harder for minority populations to purchase homes.

6.3 Neighborhoods

Most urban residential space is organized into neighborhoods, which are central institutions in everyday life and the reproduction of social networks (Fig. 6.6). They are the basis of many local friendships, support systems, and information sharing. Because they are both objective and subjective entities, neighborhoods may provide for a sense of community. They are neither reducible to, nor independent of, their residents' preferences and outlooks. Neighborhoods are critically important to the socialization of children; for smaller children, their world often is equated with their neighborhood.

Neighborhoods reflect the process by which people of similar socioeconomic status locate near one another. People with similar occupations and incomes often have similar residential

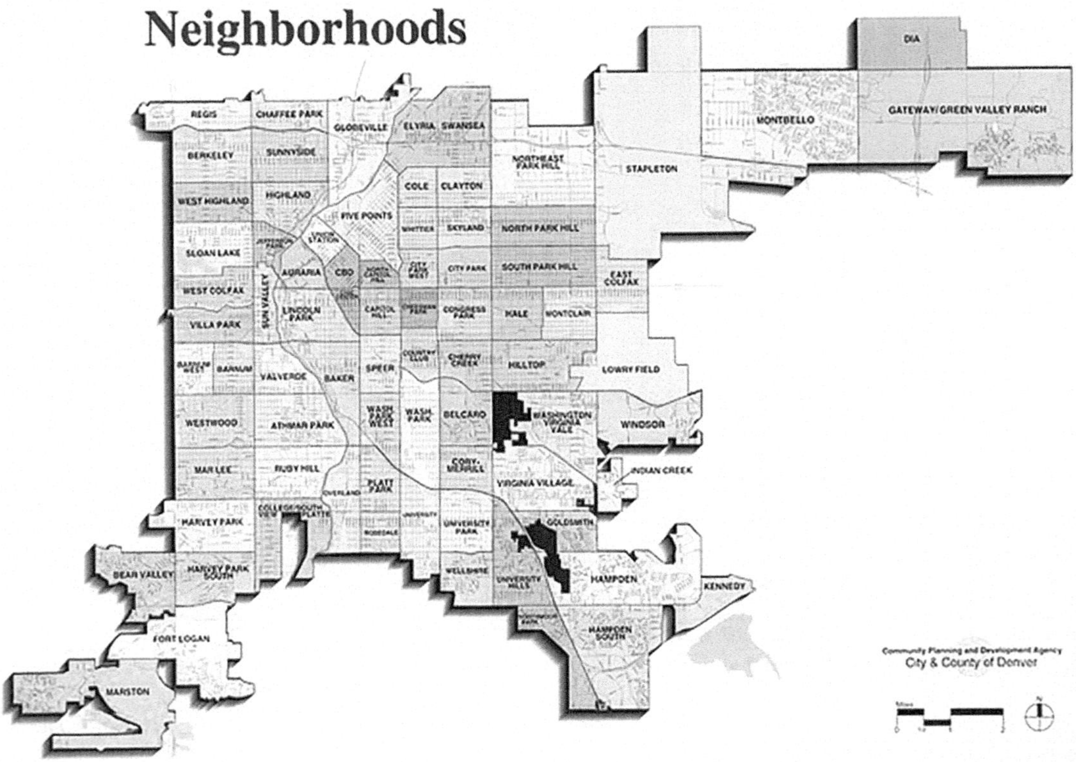

Figure 6.6

These neighborhoods in Denver reflect how residential space is organized into areas of relative homogeneity. Map lines delineate study areas and do not necessarily depict accepted national boundaries.

(https://commons.wikimedia.org/wiki/File:Neighborhoods_Map_of_City_and_County_of_Denver_-_DPLA_-_801b633fa5d3d454b323c6607e871cdf.jpg.)

preferences. Neighborhoods thus vary greatly in size and composition, ranging from impoverished locales to areas of the very wealthy (Fig. 6.7). Neighborhoods thus reflect class relations and the urban division of labor. Moreover, many neighborhoods are heavily populated by people of the same or similar ethnicity. This phenomenon was noted by the Chicago School when it observed Little Italy or Germantowns; today we witness it in the form of Chinatowns, Little Tokyo, Koreatowns, or Latino barrios. When racism was more publicly expressed in the past, neighborhoods also reflected discriminatory attempts to keep ethnic minority groups out of a given area.

Neighborhoods are not isolated entities, but are deeply shaped by the urban economy and the state. The federal, state, and municipal governments shape neighborhoods by building roads, mass transit systems, and zoning ordinances (Chapter 6).

Figure 6.7
This neighborhood in Bay City, Michigan reflects high incomes and housing prices.
(https://commons.wikimedia.org/wiki/File:Center_Avenue_Neighborhood_Residential_District_Bay_City_MI_C.JPG.)

When neighborhoods are cohesive and well-organized, they may form neighborhood associations, which help to address local communal problems. Neighborhood associations may monitor traffic volumes and press the city for speed bumps or lower speed limits or advocate for street lights. Neighborhood watch associations keep tabs on intruders and potential criminals. They may organize to voice complaints in front of zoning boards or city councils to protest things such as vacation rentals in their area. Neighborhoods are thus a primary arena in which people act as citizens, not simply consumers.

A common way in which neighborhoods organize politically is the Not in My Backyard (NIMBY) syndrome. NIMBY movements oppose locally unwanted land uses, which may include a homeless shelter, a halfway house for recovering addicts, hazardous waste treatment facilities, a land dump, or sports stadium. NIMBY movements arise when the benefits of a proposed land use stretch across the city but the costs are locally concentrated. Costs may include crime, noise, or higher volumes of traffic. When NIMBY movements begin, even mild-mannered home owners may become avid political activists defending their neighborhood's peace, safety, and property values.

Neighborhoods are never static but are constantly undergoing change. Local residents may age and die off or move to other areas. New residents may include immigrants. Shocks to municipal economies may lead to depressed housing prices and a mass exodus. Over time, technological changes have also altered the fabric of neighborhoods. Prior to the mass adoption of the automobile and telephone in the 1920s, many urban neighborhoods reflected dense networks of social relations based on face-to-face contact; many people spent their free time hanging out on the street. Today, the widespread use of air conditioning, the Internet, television, and video games has led more people to spend time inside, a hyperindividualistic situation that does not lend itself to communal bonds. The commodification of household chores (e.g., eating out) helped to undermine the stability of neighborhood ties. Finally, many inner city neighborhoods have been rocked by gentrification, the movement of relatively affluent people into once low-income communities.

6.4 Suburbanization

A major facet of urban residential space is suburbanization, the growth of areas peripheral to the urban core. Suburbanization is important because it reveals much about the urban division of labor as well as the housing preferences of large numbers of people. Suburbanization has grown steadily over the years, so that today a plurality of Americans live in suburbs.

There are several drivers of suburbanization that combine to create rapid growth on the urban periphery. Among these are: (1) Changes in transportation (e.g., streetcars, the automobile); (2) Changes in production systems (e.g., the movement of firms to the urban fringe); (3) Demographic changes (e.g., the birth of the baby boom generation in the 1950s and 1960s, as well as immigration); and (4) Government policies (e.g., the construction of highways and the federal deduction of interest on mortgages). In short, suburbanization is an economic, demographic, and cultural phenomenon simultaneously.

While suburbs are often thought to be a recent phenomenon, in fact they have a long history in the United States dating back to the late 19th century. Early suburbs were often places for the wealthy to escape the ills of the inner city such as poverty, congestion, and air pollution; a rich family might own a townhouse in the urban core and another on the periphery. Brooklyn, New York, for example, started as a suburb of Manhattan; similarly, Oak Park, near Chicago, began in 1873. Homes in these places were often largely and stately. Early suburbs thus had very few connections to the inner city, although a few people could commute on the horse-drawn omnibus. Cities in the 19th century often witnessed immigrants crowded in tenement buildings in the city and native-born residents moving to the urban edge.

Suburbs began to grow rapidly during the period of the 1880s to the 1920s under the impetus of the electric streetcar or trolley (Fig. 6.8); indeed, many were called "streetcar suburbs." Streetcars greatly lowered the time and cost to move between suburbs and the central city, allowing the middle class to move into peripheral areas. As a result, suburbs began to grow quickly in size and their interactions with the central city rose rapidly. Homes were often bought through catalogs, and developers subsidized transportation to increase property values. Most people still went to the CBD for shopping and entertainment. However, the intersections of streetcars and streets saw the rise of small stores, giving suburbs their first commercial

Figure 6.8
Streetcars in 1891.
(https://commons.wikimedia.org/wiki/File:The_street_railway_review_(1891)_(14572308857).jpg.)

functions. As suburbs grew, they were often annexed by inner cities, which sought political control in order to reap the property taxes from such areas.

The growth of suburbs in the late 19th and early 20th centuries also reflected changes in the national economy and the division of labor. Mass production became the norm, generating large numbers of well-paying jobs. With higher incomes and social mobility, many people sought to live in suburban areas. Immigrants flooded into the United States, mostly from southern and Eastern Europe, although with lower incomes they tended to be concentrated in the inner city.

In the 1920s, a period of intense urbanization, automobile ownership became widespread among the middle class (Fig. 6.9). Streetcar companies began to fail en masse as Americans took to their cars. Some cities, such as Los Angles, had light rail systems that were purchased and dismantled by automobile companies. Accordingly, suburbs witnessed a massive round of growth. Indeed, in many ways suburbs are the child of the automobile, which offered unprecedented convenience. The automobile greatly expanded the commuting range of workers, and the supply of land, and thus housing, rose exponentially. Originally, however, only men drove in large numbers. Developers would build entire communities with increasingly standardized housing styles. With rising incomes, more people could afford

Figure 6.9
Suburbs in the 1920s.
(From Museum of History and Industry, https://commons.wikimedia.org/wiki/File:New_homes_in_Laurelhurst,_ca_1925_(MOHAI_5785).jpg.)

cars, and many began to move to the suburbs *en masse*. Home ownership rates rose from 40% in 1950 to 65% in 1970, and the single-family home became the norm for the growing middle class, the symbol of realizing the American dream. Daily commuting between the suburbs and the CBD became a reality for millions of people. Suburbs thus lost their earlier upper-class status and became middle-class retreats. Many suburbs incorporated as legally separate cities to protect their property values, and taxes, from central city governments.

In the 1950s, during the post-World War II economic boom, suburbs exploded yet again. Following a prolonged period of minimal construction during the Great Depression and World War II, the demand for housing was enormous. This period also witnessed low unemployment and rapidly rising incomes. Housing starts—the supply of new housing—jumped dramatically. This process was greatly expedited by the federal government, including the GI Bill, which allowed returning veterans to attend college, and the Veterans Administration and Federal Housing Administration, which guaranteed builders that veterans could buy housing at rates lower than rental costs. By 1960, the population of the suburbs exceeded that of central cities.

One of the pioneers in the production of mass-produced housing was William Levitt and his company, Levitt and Sons (Chapter 3). They perfected the art of quickly building small, cheap, prefabricated houses, erecting an entire home in a day. The result was several communities termed Levittowns in New York, Pennsylvania, New Jersey, and elsewhere (Fig. 6.10).

It is worth stressing that suburbs during this period were almost entire populated by White people in what is popularly called "White flight." For example, Levitt refused to sell houses to people of color. In addition, ethnic minorities often had lower incomes, which made suburban housing unaffordable. Mortgage companies, or Savings and Loans, often refused to extend mortgages even to qualified people of color. Many suburbs adopted measures to keep minority groups out, including minimum lot size zoning, which specifies that homes must sit on large lots (e.g., ½ acre or more) and restrictive covenants, which prohibited owners from selling to

Figure 6.10
Levittown, Pennsylvania.
(https://commons.wikimedia.org/wiki/File:LevittownPA.jpg.)

ethnic minority groups. In short, suburbanization was deeply racialized. The result was that today, American metropolitan areas often consist of a central city with large minority populations and an archipelago of predominantly White suburbs. Los Angeles offers one such example out of many (Fig. 6.11).

Figure 6.11
Ethnic composition of the Los Angeles metropolitan region (Whites in red, Blacks in blue, Latinos in orange). Map lines delineate study areas and do not necessarily depict accepted national boundaries.
(https://commons.wikimedia.org/wiki/File:Race_and_ethnicity_2010-_Los_Angeles_(5560490330).png.)

Automobile ownership played a huge role in the growth of postwar suburbs. As incomes rose, more people could afford cars and learned to drive, including, for the first time, large numbers of women. Indeed, the single-family home, suburbs, and car ownership were all facets of one overarching social and spatial system. The Federal Interstate Highway system was instrumental in this regard, as highways greatly expanded accessibility to and from the suburbs as well as interactions among cities. The single largest project ever undertaken by the federal government, it took over 20 years to complete and has had enormous and enduring impacts on the country's cities and economy.

The explanations for suburbanization include both structural factors on the one hand and behavioral ones on the other. Structurally, historians note that the period after World War II was unique in many ways: the United States emerged almost alone in being unscathed from the war; its allies and former enemies such as Germany and Japan lay in smoking ruins. This afforded American corporations great opportunities for investments and profits. Economic growth led to unprecedented social mobility, with millions being pulled out of poverty. Home ownership rates rose steadily. Moreover, this period was also one of enormous industrial decentralization, with many companies leaving inner cities for cheaper land in the suburbs, a process expedited by the government through the interstate highway system, the GI Bill, and the deduction of interest from mortgages on federal income taxes. The decentralization of jobs was manifested in the growth of suburban industrial and office parks. In short, suburbanization cannot be comprehended without referencing labor markets and the state.

Behaviorally, suburbanization reflected not only rapidly rising incomes but deep-seated traits in American culture that linked home ownership to upward social mobility and the escape from the working class. Economists often note that most people have an income-elastic demand for space, that is, their preference for larger homes and lots increases rapidly with income. Most people want more space when their incomes rise, especially if they have children. Notably, during the postwar boom, the baby boom generation was born. Suburbs offered bountiful housing for those who could afford it, as well as good schools, peace, quiet, and relatively low crime rates. Thus, suburbanization was linked to a broader ideology focused on the nuclear family, home ownership, racial homogeneity, and "law and order."

Because suburbs are generally wealthier than inner cities, they have property tax bases that allow them to fund municipal services well. For this reason, suburban schools are often better than ones elsewhere, with smaller class sizes and more experienced teachers. In this way social inequality is reproduced through the education system.

Today, suburbs comprise a surprisingly diverse set of neighborhoods. Some are older and face similar problems to those found in central cities, including dilapidated housing, crime, and trash. Some suburbs, or "ethnoburbs," are populated by immigrants and native-born ethnic minorities, especially ones who have been in the country for years and attained a measure of disposable income. Yet other suburbs have acquired commercial functions, with downtown areas, skyscrapers, and retail complexes that rival those of the urban core, in contrast to the simplistic view that they are all bedroom communities. Indeed, most commuting today is among surburbs, not between suburbs and downtown. Suburbs thus helped to bring into being the low-density, polynucleated urban areas that we commonly see today.

Suburbanization had massive social and spatial consequences for urban cores and peripheries alike. The enormous amount of construction in the suburbs—of housing, roads, shopping malls, and office parks—covered vast amounts of agricultural land, leading to rural-to-urban land conversion. This diminished the supply of prime agricultural land, typically flat and well watered. Traffic and urban sprawl led to congestion, the destruction of wildlife habitats, and the loss of farmland. These processes also caused a deterioration of air and water quality. As the middle class moved to the suburbs, central cities saw their property tax base deteriorate, which eroded their capacity to provide public services. Poor public services, especially schools, in turn drove the middle class away, creating a self-reinforcing cycle. Retail functions in central cities also declined; some department stores closed their flagships in central cities and opened outlets in suburban malls. In this light, suburbanization and central city decline should be seen as two sides of one coin—the dual-faced nature of uneven spatial development.

Perhaps the most exclusionary space associated with suburbanization is the gated community, residential areas in which residents strictly control who may enter, often with guards (Fig. 6.12). They may be surrounded by a wall. Gated communities are typically wealthy, and offer abundant community services, including swimming pools, golf courses, and security patrols and surveillance. They tend to be homogenous in terms of ethnicity and class. The local home owners' association is very active. Such communities tend to be tight-knit, with dense webs of personal ties. Once uniquely American, gated communities have spread to many countries in the world, including China, Latin America, and parts of Africa. Critics argue that gated communities are exclusionary and erode the social capital of the city, including public life, and that they may encourage a paranoid groupthink.

On the outermost fringes of urban areas one often finds exurbs or "edge cities," far-flung suburbs that are quasirural in nature (Fig. 6.13). They may be 30 to 60 miles away from the central business district. Exurbs have grown rapidly in recent years as large numbers of people seek a combination of rural amenities and access to suburban services. Their residents tend to be above average in income. In such areas, lot sizes tend to be large and population densities are very low. Commutes are typically quite long, although the ability to work from home using the Internet has increased their appeal. Often they are unincorporated, meaning residents do not pay municipal property taxes or enjoy municipal services. Public transportation is unavailable, making the automobile a necessity, and they tend to be less walkable than suburbs. They tend to lack the commercial or retail functions found in many other suburban areas. About 6% of the United States lives in exurbs.

Figure 6.12
Gated community in Saskatoon, Canada.
(https://commons.wikimedia.org/wiki/File:Saskatoon_gated_community.JPG.)

6.5 Gentrification

Another major force shaping Western cities is gentrification, a term coined by the urban planner Ruth Glass as the replacement of working-class people by a gentry, that is, a relatively wealthy elite. The movement of high-income people into modest neighborhoods changes not only the area's population composition, but real estate values and local retailing. Gentrification is sometimes portrayed as the opposite of suburbanization, but that is a simplistic interpretation; moreover, gentrification is much smaller than suburbanization in terms of the numbers of people involved.

Gentrification takes different forms in different contexts. In some cases it may see "yuppies," or young urban professionals, purchase run-down properties in historic neighborhoods and fix them up. Artists may purchase lofts that were formerly parts of warehouses, which were once cheap but are now high-priced. In other cases, developers may build condominium towers catering to the wealthy (Fig. 6.4). As moneyed professionals move in, they support ancillary services like restaurants, galleries, bars, boutiques, bistros, and personal services. Finally, commercial developers may then enter an area to build convention centers, waterfront

Figure 6.13
Aerial view of Falcon, Colorado, illustrates the quasirural nature of exurbs.
(https://commons.wikimedia.org/wiki/File:Falcon,_Colorado_(14017129080).jpg.)

properties, sports arenas, office towers, and recreational complexes, often with the active encouragement (and subsidies) of municipal governments. Baltimore's Inner Harbor is a good example (Fig. 6.14). Many urban governments favor gentrification because the gentry place few demands on public services (e.g., many are childless and thus don't rely on public schools) and the process improves property values and, thus, taxes. Gentrification is thus often accompanied by municipal rezoning, or changes in the legal restrictions on land use.

Gentrification began in the 1970s in deindustrialized cities such as Philadelphia, which had been rocked by decades of deindustrialization, job loss, and urban decay. Abandoned by capital, many urban cores had abandoned buildings and were ripe for redevelopment. Many such cities had acquired large stocks of dilapidated housing acquired when property owners failed to pay property taxes and were eager to turn them into productive units. For example, Baltimore began selling run-down units for one dollar. From northeastern cities, gentrification

Figure 6.14
Baltimore's inner harbor.
(https://commons.wikimedia.org/wiki/File:Inner_Harbor_Aerial_2022.jpg.)

spread rapidly to change the core of almost every large city, and even many smaller ones, becoming a ubiquitous force across the American urban landscape.

The movement of relatively well-to-do people into urban cores was initially likened to 19th century homesteading on the Midwestern prairies, and many yuppies saw themselves as living on an "urban frontier" or "urban jungle." This type of view overlooks the fact that gentrified areas had people living in them, and the frontier analogy dehumanizes them, reducing them to savages.

The roots of gentrification are complex, and may be interpreted in two different ways. First, it is no accident that the widespread revival of residential space in central cities in the late 20th century occurred precisely as corporate America reclaimed the cores of metropolitan areas. This shift reflected in particular the massive growth of financial and producer services, which agglomerate in downtown areas (Chapter 4), thus making centralized locations optimal. Many workers in such industries fall into the "creative class" of professionals in skilled, white-collar occupations who earn respectable salaries. In this reading, gentrification is symptomatic of the emerging globalized, digitized urban division of labor that emerged in the 1980s and 1990s.

One variant of the structuralist interpretation of gentrification is the rent gap, a term coined by the geographer Neil Smith. The rent gap refers to the difference between the potential rents that landlords may receive for urban properties and the actual rents that they receive. When this gap becomes sufficiently large, as when inner city neighborhoods experience decay and job loss, the incentive for developers to move in and rebuild an areas become immense.

A different perspective holds that gentrification is the result of changes in residential preferences, a view that stresses the behavioral and psychological dimensions rather than the world of work. This perspective emphasizes that many people have few or no children and exhibit a strong preference for urban living, particularly in core areas with ready access to restaurants and night life: the "bright lights, big city" phenomenon. Wealthy, well-educated professionals in two-income households with significant disposable incomes may pay a premium for such units. Such people tend to like historic neighborhoods with their unique architecture. They often celebrate the diversity of urban life even as gentrification diminishes it. In this light, gentrification is primarily a cultural and demographic phenomenon.

Gentrification has not unfolded without significant social costs. First, the gentrification of neighborhoods inevitably results in rapidly rising property values, and thus rents. Long-time residents may find it increasingly difficult to live in such areas, and be forced out. Small "mom and pop" stores tend to fold, replaced by large retail chain stores. Rising rents lead to shortages of affordable housing. Indeed, increases in rents in gentrifying areas contribute to the crisis of homelessness in many urban areas. Moreover, gentrification frequently has a racial overtone: the "yuppies" moving into gentrified areas tend to be overwhelmingly White, whereas the local residents are frequently people of color (Fig. 6.15). Typically, when neighborhoods occupied by underrepresented communities gentrify, they start to become increasingly White. For these reasons, gentrification is often actively opposed by local residents (Fig. 6.16), who correctly view it as a threat to the fabric of their neighborhood. Some portray it as "urban colonialism" or class warfare and bemoan the inequality that results. In short, gentrification is a class-based, racialized, political and cultural phenomenon.

Once a strictly American phenomenon, gentrification has spread throughout the world. Propelled by the surge in producer services, including finance, it has become a potent force reshaping cities in many continents. In London, the famous Docklands is a prime example of corporate-led gentrification. Cities in Canada, Europe, and China have been similarly affected. Even in the developing world, cities such as Mexico City exhibit signs of it, as do some in India.

6.6 The Crisis of the Urban Ghetto

One of the more odious aspects of American cities is the formation of urban ghettos, impoverished neighborhoods located near the inner city. Ghettos are not new: the term originally applied to areas in European cities to which Jews were confined in the medieval era. The first ghetto was in Venice, Italy. However, the term has become generalized. Although American ghettos were populated by immigrants from southern and Eastern Europe in the late 19th and early 20th centuries, ghettos today are primarily inhabited by African Americans and Latinos (in which case the term "barrio" is more appropriate).

Figure 6.15

Gentrification architecture in Beford-Stuyvesant, an historically African American community in Brooklyn, New York.

(File: Gentrification architecture in Bedford Stuyvesant.jpg—Wikimedia Commons.)

Figure 6.16

Graffiti protesting gentrification.

(https://commons.wikimedia.org/wiki/File:STOP_GENTRIFICATION_graffiti_in_Turin.jpg.)

The creation of large African American communities in American cities began in earnest shortly after World War I, when large numbers of returning Black veterans began to settle down and look for jobs in manufacturing. In the 1920s and for several more decades, streams of African Americans began to leave the South in an event often called the Great Migration (Fig. 6.17). This process reflected the confluence of several factors: the mechanization of agriculture in the South; the abundance of jobs in growing Manufacturing Belt cities; and Congressional limitations on immigration, which restricted the supply of labor. As a result, Black communities exploded in size. Sustained by relatively high wages, any Black people left poverty and entered the middle-class; stable communities with high rates of home ownership were evident in cities such as Boston, New York, Chicago, Pittsburgh, Philadelphia, and Cleveland. During this period, Harlem, in northern Manhattan, became the cultural and spiritual center of African American life, an event known as the Harlem Renaissance.

This situation did not last long, however. After World War II, when waves of White flight saw the White middle class evacuate central cities, large urban areas began to experience growing economic and social problems. African Americans never enjoyed the same levels of growth during the postwar boom as did Whites. Large cities began to lose people, and because Whites

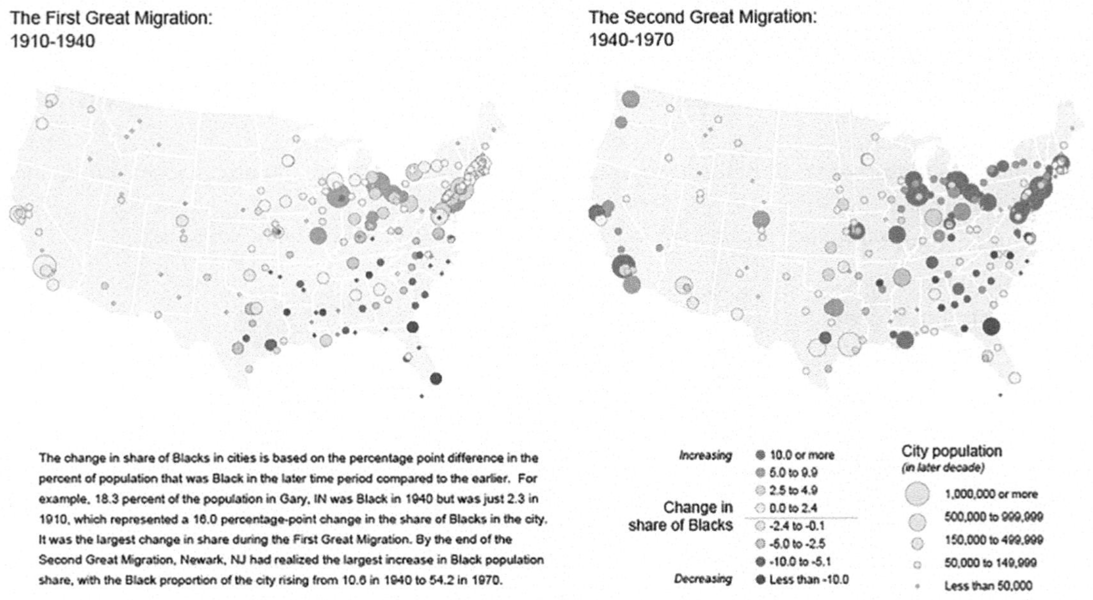

Figure 6.17

The Great Migration saw waves of African Americans leave the South and move to industrial cities in the Northeast and Midwest. Map lines delineate study areas and do not necessarily depict accepted national boundaries.
(https://commons.wikimedia.org/wiki/File:GreatMigration1910to1970-UrbanPopulation.png.)

were more likely to leave, many inner cities became heavily populated by Black people. In 1990, for example, Detroit was 75% Black, followed by Washington, DC (65%), Atlanta (67%), Memphis (64%), New Orleans (62%), and Baltimore (59%). One unanticipated outcome was to make African Americans a powerful voting bloc in these cities, which often had Black mayors.

By the 1970s, deindustrialization became a key motor of this process, leading to plant closures and growing unemployment. African Americans were disproportionately concentrated in manufacturing jobs, and were thus especially vulnerable when industrial firms began to close down or move overseas. Often suffering from low levels of education and institutionalized racism, many Black people lacked reasonably paying alternative jobs, and sank into poverty. Black unemployment rates are typically two or three times that of White people. The middle-class communities that arose in the 1920s began to turn into ghettos. Today, Black incomes are only about 70% of that of Whites, and Black wealth (accumulated assets) is only 5% of that of Whites, largely due to lower levels of home ownership.

Black poverty and low levels of home ownership were compounded by a common practice known as redlining, which is the formal or informal denial of mortgage applications by lending institutions to people of color (Fig. 6.18). Redlining was common in the early 20th century and had severe impacts on urban residential landscapes—a prime example of racism embedded in the urban landscape. Even qualified Black buyers were often turned down. Lending institutions such as Savings and Loans denied that they engaged in racism, arguing that by denying loans to lower-income people they were protecting their stockholders. In the 1960s, as the Civil Rights movement grew, redlining was rendered illegal by the Fair Housing Act of 1968. However, it continued to be practiced by many firms illicitly.

Affordable housing is always in short supply, but the housing crisis hit Black communities particularly hard. Many landlords in ghettos refused to invest in their own properties, such as fixing plumbing and heating issues. Some even burned their own buildings down to collect insurance money. The result was a deterioration in the housing stock that left many lower-income people living in deplorable conditions. In the 1950s and 1960s, this problem was made even worse by "urban renewal" (Chapter 8), a federal program ostensibly designed to modernize urban cores. Known informally as "Negro removal," urban renewal witnessed the destruction of large numbers of Black housing projects and apartment buildings to make way for freeway on-ramps, shopping centers, sports arenas, convention centers, and the like. Urban renewal was essentially government-backed slum clearance. Hundreds of thousands of Black people were displaced by this process. Mass unhappiness with housing was a major incentive behind the eruption of many Black urban communities in the 1960s.

The result of decades of White flight and Black impoverishment is that the typical pattern of American cities is to have a large community of African Americans near the urban core and

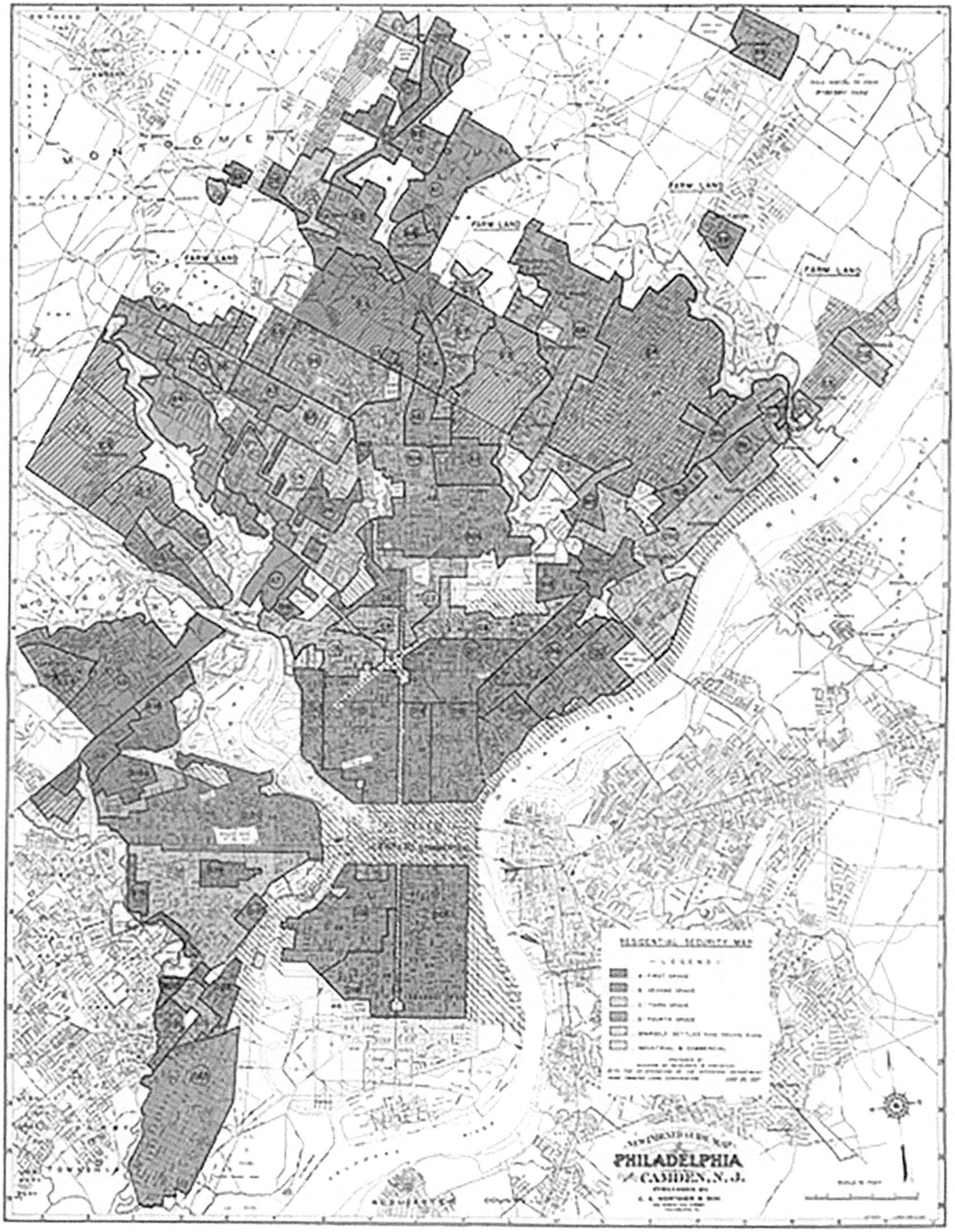

Figure 6.18

An example of redlining in Philadelphia in 1937. Map lines delineate study areas and do not necessarily depict accepted national boundaries.

(https://commons.wikimedia.org/wiki/File:Home_Owners%27_Loan_Corporation_Philadelphia_redlining_map.jpg.)

White suburbs surrounding them (Fig. 6.19), a configuration repeated over and over again in hundreds of cities. In this way, uneven urban spatial development is racially polarized.

In the face of long-term capital disinvestment, mounting poverty, and high unemployment, the quality of life in many ghetto areas began to deteriorate. Public services, including education, transportation, health care, and trash removal began to falter. Often gangs arose given the

Figure 6.19
Distribution of Blacks and Whites in Cleveland, 2010. Map lines delineate study areas and do not necessarily depict accepted national boundaries.
(US Census Bureau [https://www.census.gov/hhes/www/housing/housing_patterns/pdf/Fig5-3.pdf].)

inadequate police protection cities offered. Graffiti was not removed. Abandoned cars were never picked up.

Today, many ghetto residents face bleak prospects (Fig. 6.20). Such areas are often trapped in unskilled, low-wage economies, offering little room for social mobility. Skilled white-collar jobs in the producer services that agglomerate in the central business district are often out of reach, partly due to poor public education systems. High school dropout rates are above average (but have come down). Many jobs, such as fast food and security, pay near minimum wage and offer few benefits. Some find employment in blue-collar occupations (e.g., construction, auto body shops, plumbing, health care). A disproportionate number join the military. Crime rates are often high (10% of Black men are in prison or on probation). Drug use is often widespread, gangs are locally powerful, and many suffer from police brutality. Public hospitals are overcrowded and schools are underfunded, with large classes and ill-prepared teachers. Gentrification of the inner city has made affordable housing scarce. Moreover, these problems are compounded by cutbacks in government services, such as housing assistance, that often result in homelessness. Threats of eviction hang over many people. There are tens of millions of families living paycheck-to-paycheck, living in a state of quiet desperation and severe financial uncertainty.

The seriousness of ghettos sparked an enormous debate as to the causes, with serious implications for urban policy. Essentially this debate was about culture versus jobs. A conservative argument held that the social programs of the 1960s, such as President Johnson's War on Poverty, led to a "culture of poverty" that included dependency on "welfare" payments. In this view, social programs create a disincentive for lower-income people to find

Figure 6.20
Two men in the Chicago ghetto.
(https://commons.wikimedia.org/wiki/File:Chicago_ghetto.jpg.)

work. "Freeing" people from the "shackles" of welfare was thus held to stimulate job-seeking and an escape from poverty. The result was the creation of an "underclass" (the term is questionable) that is not only unemployed, but unemployable.

Critics allege that this view blames poor people for being poor. These people are politically powerless, and making them scapegoats is a long-standing strategy for many politicians. The argument that people living in poverty do not wish to work is belied by the long lists of applications in ghetto areas even for menial jobs such as fast food. Liberals argue that the conservative focus ignores the macroeconomic context of American cities and the structural forces behind ghetto formation, including deindustrialization and persistently high levels of unemployment.

Famed sociologist William Julius Wilson took this line of thought farther, noting that high levels of male unemployment led to a shortage of marriageable partners. As a result, many Black children (70%) are born out of wedlock, that is, they are raised by single mothers. Ethically there is nothing wrong with this; many women are better off without unproductive men. But financially this is disastrous; the bulk of people living in poverty are single mothers with children. Historically this was not the case: Black families survived centuries of oppression under slavery. But deindustrialization did what slavery could not, fragment the Black family.

Ghetto problems can be softened, if not solved, by aggressive public policies to help low-income people. Steps include a higher minimum wage; drug counseling and rehabilitation programs; needle exchanges for IV drug users; tax credits and subsidies for low-income housing; job training programs; improved public transportation; and neighborhood recreational facilities (e.g., basketball courts).

6.7 Homelessness

At the very bottom of the pecking order in residential space are displaced people, those who do not have a place to live. Homelessness is a serious public health and policy issue with complex roots and important consequences.

Definitions of homelessness vary, ranging from people who might spend a night or two on a park bench or shelter to those who have been living on the streets for long periods of time. By the latter definition, somewhere between 500,000 and 1 million Americans are homeless, although the population is difficult to count accurately. Most homelessness is temporary, as people float in and out of the housing market: the average duration is 6 months.

Homelessness has a distinct geography. It is, not surprisingly, concentrated in states and cities where housing is the most expensive (Fig. 6.21). California, where housing is notoriously expensive, has 12% of the US population but 50% of its homeless people. Within states,

Figure 6.21

Homelessness by State in 2020. Map lines delineate study areas and do not necessarily depict accepted national boundaries.

(https://commons.wikimedia.org/wiki/File:2020_HUD_AHAR_homelessness_map.svg.)

homelessness tends to be concentrated in large cities, including New York, Boston, Los Angeles, San Francisco, and Chicago. However, there are also smaller populations of suburban and even rural homeless people. Ironically, homelessness is worst in wealthy states, not poor ones. Even liberal states such as California that are nominally more empathetic to displaced people have very strict restrictions on new housing, which aggravates the problem.

There are several intersecting causes of homelessness. By far the most common explanation is the lack of affordable housing: most people are without permanent homes simply because they cannot pay the rent. The supply of affordable housing is also threatened by forces such as gentrification. The problem has been exacerbated by decades of budget cuts to federal assistance in housing and poverty relief. In essentially no city in the United States does earning the minimum wage allow people to rent a two-bedroom apartment. Vulnerable pepole are often threatened with evictions before losing their homes.

To the surprise of many, roughly 20% of displaced people work full time, such as in grocery stores, nursing homes, and hospitals: they are caught between jobs that pay too little and housing that costs too much. Most are poorly educated and have few marketable skills. Homelessness undercuts the traditional assumption that having a job keeps one off the streets. These are people living paycheck-to-paycheck. For every person living on the streets, there are many others who are only one missed paycheck or rent hike from homelessness. Homelessness has grown as the cost of housing has outpaced incomes in most places. Moreover, homelessness changes with the business cycle, rising during recessions.

Other reasons include chronic drug and alcohol abuse. Some are women fleeing domestic violence, or divorced mothers whose ex-husbands do not pay child support. There are also deinstitutionalized psychiatric patients. Debates rage as to whether homeless schizophrenic people may be hospitalized against their will. (However, rates of mental illness among displaced populations are only slightly above that of the population as a whole.) Veterans are disproportionately unhoused as well. Although roughly 60% of displaced people are men, the most rapidly growing population consists of women with children. Yet others are refugees from the foster care system, including runaway teenagers. There are numerous homeless students, including some in college. This means the unhoused population is diverse and the problem cannot be remedied by simplistic one-size-fits-all policies.

Being homeless is a very hard way to live. Living on the streets is uncomfortable and often dangerous (Fig. 6.22). Many are hungry, malnourished, and perpetually stressed out. They are scorned, insulted, and shunned, reviled by the public because of their appearance and smell. Some homeless people panhandle aggressively. Without adequate access to restrooms, some engage in public urination and defecation. Some are victims of violent crimes. Winters are the hardest time for unhoused people, when many die of hypothermia. Some cities such as New York force them into shelters during the winter, even if involuntarily. Approaching the problem of homelessness requires empathy, sympathy, and understanding.

One of the most pernicious and destructive myths about homelessness is that it is voluntary. This has been a long-standing claim by forces that seek to reduce funding for displaced people, who argue that unhoused people prefer free meals in churches and food pantries rather than working for a living. For example, former President Ronald Reagan once claimed "What we have found in this country … is that the people who are sleeping on the grates, the homeless, you might say, do so by choice." This is absurd. No one chooses to be homeless. The idea has floated around since the Great Depression of the 1930s, when many claimed that "bums" refused to work. This view underlies policies such as workfare, that force unhoused people to seek jobs, even when they are scarce. Displaced people are a politically impotent population, and it is easier to blame the victims than to find solutions. Discourses that blame the unhoused, like those living in poverty, overemphasize personal behaviors and overlook the structural causes such as labor and housing markets.

When people become homeless, it is common to progress through a series of stages marked by increasing desperation. In the initial stages, they may double up with friends or family,

Figure 6.22
A homeless encampment in Portland, Oregon.
(https://commons.wikimedia.org/wiki/File:Northeast_Portland_homeless_camp_tents.jpg.)

engaging in "couch surfing." Some fall under the category of automotive homelessness, living in their cars. Later stages involve living in a homeless shelter; some are reluctant to do so out of fear of crime, or prohibitions in shelters against drug and alcohol use. The demand for beds in homeless shelters almost always exceeds the supply.

Like poverty, there are ways in which public policy can at least soften the hardest edges of homelessness. A starting point is to create more beds in shelters, which alleviates the emergency conditions that many unhoused people face. Cities may provide free rooms in cheap hotels. Stronger tenant protections, such as strengthening just-cause laws about evictions, would protect those barely able to pay rent. Housing vouchers are another step. Cities can revisit antiloitering laws, which make it illegal to sleep in parks. Raising the minimum wage would help. Cities and states can provide job training programs and job banks. Centers for drug, alcohol, and psychiatric counseling are important. People with severe mental illnesses should be reinstitutionalized. A major benefit would be to increase dramatically the supply of low-income housing. European countries also have homelessness, but not to the degree prevalent in the United States, largely because the social safety net in Europe is more generous.

6.8 Chapter Summary

Residential space reflects the fact that people live in cities, and do not simply work in them. However, housing and jobs are always intertwined: urban spaces of production and social reproduction are deeply engaged with one another.

The study of cities was launched by the hugely influential Chicago School, which gave rise to urban geography and sociology in the early 20th century. Since then, our understanding of this issue has progressed mightily. The dynamics of housing markets, which reflect complex constellations of factors ranging from demographic change to interest rates, dictate how available and affordable housing in a city will be. For people with low incomes, affordable housing is always in short supply.

Neighborhoods arise from similar residential choices made by people with similar incomes, ethnicities, and preferences. Neighborhoods are an important local context for understanding urban daily life and the socialization of children. At times they are politically active, such as in Not in My Backyard (NIMBY) movements.

Several facets of residential space reflect how changing production system dynamics and labor markets reshape urban space. For more than a century, the middle class has fled inner cities for the suburbs, where housing is in abundant supply. In the United States, suburbanization has long been racialized and became synonymous with White flight. Conversely, given the explosion of producer services in the late 20th and early 21st centuries, gentrification has seen many relatively wealthy professionals, usually White, move into working-class neighborhoods near the urban core: for some, this is a process of revitalization; for others, it symbolizes wealthy elites taking over parts of the city previously occupied by low-income communities.

The racialized dynamics of production and residential space also produced the inner city ghetto, impoverished regions of unemployed or underemployed people marginalized by the producer service economy. Most inhabitants of American ghettos are African Americans, although the growth of Latino barrios indicates that processes of marginalization and exclusion have extended to other ethnic groups as well.

Finally, this chapter addressed the question of homelessness. In the United States, one of the wealthiest economies in the world, roughly a million people lack a home. The causes of homelessness primarily lie in the fact that housing is expensive and many jobs simply do not allow people to afford a place to live. The notion that homelessness is the result of choice is simply wrong. Other factors include the deinstitutionalization of psychiatric patients, veterans who have not received sufficient care, the foster care system, and women fleeing abusive partners.

Study Questions

1. What was the Chicago School and how did they view the city?
2. What forces determine the nature of neighborhoods?
3. What are the driving factors behind suburbanization?
4. How is suburbanization racially stratified?
5. How is gentrification linked to urban labor markets? What causes it?
6. Why is gentrification often opposed by local residents?
7. Is homelessness caused by people simply choosing to go without housing? What factors are responsible?

Further Reading

Martin Brockerhoff, Neil Smith, The new urban frontier: Gentrification and the Revanchist City. Population and Development Review. 00987921. 23 (2) (1997), 441, Available from: https://doi.org/10.2307/2137563.

Chad R Farrell, Immigrant suburbanisation and the shifting geographic structure of metropolitan segregation in the United States. Urban Studies. 0042–0980. 53 (1) (2016), 57–76, Available from: https://doi.org/10.1177/0042098014558537.

M. Foscarinis, And Housing for All: The Fight to End Homelessness in America. Prometheus. (2025).

C. Hochstenbach, S. Musterd, Gentrification and the suburbanization of poverty: Changing urban geographies through boom and bust periods. Routledge, Netherlands Urban Geography. 02723638. 39 (1) (2018), 26–53, Available from: http://www.tandfonline.com/loi/rurb20?open=34&repitition=0#vol_34, https://doi.org/10.1080/02723638.2016.1276718 .

Carol E. Hoffecker, Kenneth T. Jackson, Crabgrass frontier: The suburbanization of the United States. Journal of Interdisciplinary History. 00221953. 17 (3) (1987), 691, Available from: https://doi.org/10.2307/204639.

P. Knox, S. Pinch, Urban Social Geography: An introduction Urban Social Geography: An Introduction. Taylor and Francis Inc., United States (2014), 1, Available from: https://www.routledge.com/products/9780367239312. https://doi.org/10.4324/9781315847238.

Colin McFarlane, The city as assemblage: Dwelling and urban space. Environment and Planning D: Society and Space. 0263–7758. 29 (4) (2011), 649–671, Available from: https://doi.org/10.1068/d4710.

K. Taeuber, A. Taeuber, Residential Segregation and Neighborhood Change. Transactions Publisher. (2008).

Emily Talen, Neighborhood. Oxford University Press, (2019), https://doi.org/10.1093/oso/9780190907495.001.0001.

R. Walker, R.D. Lewis, Beyond the crabgrass frontier: Industry and the spread of North American cities, 1850–1950. United States. Journal of Historical Geography. 03057488. 27(1):3–19, 2001, Available from: http://www.elsevier.com/inca/publications/store/6/2/2/8/8/0/index.htt, https://doi.org/10.1006/jhge.2000.0266.

CHAPTER 7

Environmental Dimensions of Cities

Chapter Outline

Cities do not exist in an environmental vacuum. Rather, they are always and everywhere embedded within the biophysical environment, what is commonly called "nature," including their climate, water systems, topography, soils, and flora and fauna. Energy and water flows are as critical to the existence of cities as people and transportation systems; urban areas cannot be understood without them. Cities take in water, air, materials, and energy to build the environment and allow people to eat and produce waste. This means that cities are as much environmental, ecological, and biological phenomena as they are social and economic creations.

This chapter explores the environmental dimensions of cities in several ways. It begins by describing urban climates, in particular urban heat islands. Second, it addresses air pollution, a long-standing problem of life in cities. The third part focuses on urban water systems. Fourth, it looks at waste removal. In the fifth section, it turns to green spaces and parks, noting their critical roles and uneven access to them. Sixth, it notes the process of rural-to-urban land conversion, in which agricultural areas on the urban periphery are swallowed up by housing projects and retail spaces. The next section briefly summarizes urban political ecology, which views urban environmental issues in terms of social relations such as class, gender, and ethnicity. Next it turns to urban hazards, ranging from earthquakes and hurricanes to floods. The last section focuses on climate change, the specter of which hangs over cities and is already making life difficult for large numbers of people across the planet.

Urban Geography. DOI: https://doi.org/10.1016/B978-0-443-30281-7.00007-4

7.1 Urban Climates

Daily lives in cities are shaped by the climates in which people live—the temperature, seasonal fluctuations, humidity, prevailing winds, and rainfall. Thus urban climatology is an important and complex topic. Typically studies of urban climates focus on areas beneath the urban canopy layer, or regions beneath the tallest buildings, as distinct from the urban boundary layer above them. A given city's climate is determined by its latitude, altitude, and location relative to large water bodies, prevailing wind patterns, and local topography.

In general, cities are warmer than rural areas. It is interesting to note that cities do not simply exist in their climates, they actively shape them, which separates cities from surrounding rural areas. Changes in land use affect the vegetation cover and thus transpiration, runoff from rainfall, and local precipitation levels, humidity, and temperatures. In short, cities alter their microclimates, sometimes drastically. Arguably the most obvious example of how cities change their climates is urban heat islands (Fig. 7.1). Because surface areas in cities are far more likely to be paved over than in rural areas, cities reflect more heat back into the atmosphere, a process aggravated by automobiles, power plants, and air conditioners. Streets and walls absorb heat during the day and reemit it at night; the urban climate effect is thus strongest at night. Air pollution can act like a mini-greenhouse, raising temperatures. Suspended particulate matter over cities contributes to higher rates of cloud formation and precipitation. Buildings block the wind, which lowers temperatures, although on occasion, cities may have wind tunnels. Cities are on average 1–10 degrees warmer than their surrounding rural areas. The size and intensity of urban heat islands reflects how built-up an urban area is, the types of materials used for roofs and pavements (which have varying albedos, or reflection rates), and the presence of trees, parks, and gardens. In cold climates, urban heat islands may result in warmer winter nights. However, heat islands also lead to higher rates of energy consumption (e.g., air conditioners) and may create health hazards for

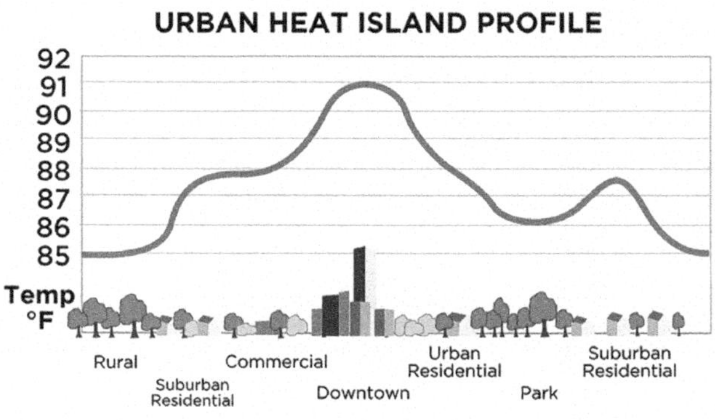

Figure 7.1
Urban heat island profile.
(https://commons.wikimedia.org/wiki/File:Urban_heat_island.svg.)

people susceptible to overheating. Heat islands can be mitigated by planting trees and vegetation and using roofing materials with a high albedo.

Trees are important to urban climates. Trees provide shade, filter the air, are homes to birds and squirrels, and their transpiration of water helps to reduce local temperatures around them. Unfortunately, tree canopies are far more likely to be found in affluent neighborhoods than in poor ones. In American cities, the wealthiest neighborhoods have 50% more tree cover than do low-income areas. This means that richer parts of the city are cooler and more comfortable than poorer parts. In this light, urban tree cover is a matter of environmental justice.

Urban temperatures can become serious hazards during heat waves, when temperatures are highly elevated for prolonged periods of time. Urban heat waves can be devastating, killing thousands of people at one time. In 1972, a heat wave in New York killed 900 people. In 2003 a European heat wave killed an estimated 15,000 people. Heat waves are most likely to affect older people and those with fewer financial resources, who may not have access to air conditioning. Heat waves have been greatly aggravated by climate change, which has driven temperatures in many areas to new heights. Already some cities in South Asia have been heavily affected by heat waves that drive temperatures up to intolerably high levels.

Some cities attempt to mitigate the effects of heat islands, heat waves, and climate change by building "green infrastructures." These include bioengineered buildings that provide ecosystem services like microclimate regulation, by building gardens, parks, marshes, and green spaces. Some strive for passive cooling using inventive architectural structures, including white roofs that reflect more sunlight, as well as permeable pavements. Other strategies include reducing the amount of impermeable land, which decreases runoff and hence flooding, water capture, storm surge protection, retrofitting stormwater reservoirs to hold water longer, and tree-planting programs. Green infrastructures improve urban resilience and cities' ability to withstand the effects of climate change. For this reason, many cities have green infrastructure programs.

7.2 Urban Air Pollution

The air that people breathe is another environmental dimension of cities that deserves scrutiny. Cities have long been notorious for having poor air. During the Roman Empire, residents complained about the city's poor air. Medieval cities were filthy (Chapter 3), and even then residents complained about the smell of garbage, pigsties, and horse manure. Many people burned wood indoors, leading to respiratory problems. The invention of the chimney was a significant step forward in having cleaner air indoors.

During the Industrial Revolution, which greatly accelerated the magnitude of air pollution, smoke from factories was widespread, as was the burning of coal in homes and industrial plants, and at times it killed thousands. The use of coal for factories, railroads, and homes (for heating and cooking) generated enormous amounts of soot that was spewed into the air. Essentially all cities had air quality problems, leading to widespread respiratory diseases such as bronchitis and asthma, and thus high mortality rates, lowering life expectancies. By the late 19th century many cities had initiated restrictions to try to limit coal emissions. Even so, at times air pollution was regarded as a sign of wealth.

The widespread adoption of the automobile in the early 20th century marked a new chapter in urban air pollution. Exhaust from cars came to be known as smog, a portmanteau of "smoke" and "fog." Cars emit roughly 5000 compounds that may be considered pollution, many of them toxic and some carcinogenic. The most common are sulfur dioxide (SO_2), nitrogen dioxide (NO_2), carbon monoxide (CO), and ozone (O_3). Air pollution has long been associated with respiratory diseases and asthma. Some cities, such as Los Angeles, became infamous for their high degrees of pollution in the 1960s and 1970s, which spurred the city, and the state of California, to adopt strict pollution controls. Today, in most cities in the developed world, urban air quality is closely monitored as a public health concern. On days with high levels of pollution, people may be advised to stay indoors.

A particularly dangerous form of urban air pollution occurs during temperature inversions, in which a layer of cooler air lies atop a layer of warmer air, an inverse of the normal pattern. Inversions vary in intensity, but can trap smog at deadly levels. An inversion caused the infamous "London Fog" of 1952, which killed between 4000 and 12,000 people in four days. Temperature inversions occur regularly over many large cities, such as Los Angeles and Shanghai.

The impacts of air pollution are not equally felt by all social groups. In the United States, people of color are far more likely to be negatively affected than are Whites, which indicates that air pollution is a matter of social justice. Asthma rates among African American children, for example, are much higher than among White children. The pioneering work of Laura Pulido demonstrated that in Los Angeles, the distribution of air pollution was largely determined by the residential locations of ethnic minorities such as African Americans and Latinos.

Essentially all countries in the developed world have adopted measures to improve urban air quality. Carburetors were replaced by fuel injection systems, and many areas require inspections to test the exhaust from automobiles. Many cars had catalytic controls installed. As a result, air quality today tends to be considerably better than it was two generations ago. The encouragement of green spaces helps to reduce air pollution; trees absorb large quantities of carbon dioxide, the primary gas implicated in climate change.

In the developing world, however, urban air quality is a pressing health concern (Fig. 7.2). Worldwide, more than one million people die annually from urban air pollution. Many densely populated cities have very bad air. India, for example, has some of the most polluted cities in the world. Air pollution in the developing world has several origins, including

Figure 7.2
Air pollution in Taiwan.
(https://commons.wikimedia.org/wiki/File:Air-pollution-taiwan.JPG.)

factories and automobiles, but another cause is the burning of vegetation on fields after the harvest, which releases vast amounts of smoke. This is a major reason why many Indian cities have such poor air quality. The worst air pollution occurs when extremely small particulates are suspended in the atmosphere; these penetrate deep into the alveoli during respiration, and can cause slow suffocation.

7.3 Urban Water Systems

Obviously, people in cities need water. This constraint was one reason the world's earliest cities were often located near rivers (Chapter 2). The history and geography of urbanization has also been marked by attempts to bring water to communities by whatever means possible. Mohenjo-Daro, on the Indus River, had sophisticated water delivery systems. The Romans constructed large aqueducts, using gravity to bring water from remote hills to their cities (Chapter 2). When available, many people used wells and cisterns.

In early modern cities lacking adequate water infrastructures, clean water was difficult to obtain. Polluted water supplies, in addition to crowded residential spaces and streets and inadequate removal of garbage, contributed mightily to the spread of diseases such as cholera, from which countless people died. Those who could afford it drank slightly alcoholic fermented fruit juices, which were at least sterile.

In the late 19th century, many European cities began to build centralized water systems, including aqueducts and pipelines, as well as sewer lines. The introduction of indoor plumbing greatly increased the urban demand for water, including faucets and toilets, and improved sanitation conditions immensely. The public infrastructure encouraged people to stop using private water sources such as wells and rain barrels.

In the early 20th century in both Europe and North America, state and local governments began to build dams and reservoirs to ensure that their urban populations had a steady and reliable source of clean water. Some floated bonds to raise funds in order to do so. All cities began construction of underground water lines to provide water. The availability of clean water had dramatic impacts on urban health, reducing many waterborne diseases like malaria cholera and typhoid fever and in the process, raising life expectancies. Along with water lines, cities also built water treatment plants to process sewage and stormwater runoff.

Vast quantities of clean water are necessary to support the daily needs of residents in large cities. New York City alone uses one billion gallons of water each day. With an aging system of pipes and aqueducts, the water system of New York leaks thousands of gallons per day. Identifying and fixing leaks in urban water systems is an important task involving engineers and hydrologists.

Many places experience water shortages, leading cities to implement restrictions on water use. These may take the form of asking people to take shorter showers, prohibiting the watering of lawns and washing of cars, or even filling swimming pools. Phoenix, Arizona, situated in a dry, hot desert, is a case in point; the city, as well as agriculture in southern California, have bled the Colorado River dry, and its outlying areas have dire water problems. In such cases cities often encourage residents to plant native xerophytes, or vegetation that evolved in dry climates and requires less water.

In the developing world, clean water can be difficult to obtain. One-quarter of the world's people do not have regular access to sanitary water. In many countries, people living in poverty rely on communal wells to obtain water; fetching water is typically considered women's work, and many women spend long hours walking to wells, waiting, and carrying heavy containers home. In times of severe water shortages, they may turn to private water vendors, who sell water at exorbitant rates. In short, water is as much a sociopolitical phenomenon as it is a "natural" or chemical one.

7.4 Waste Removal

Cities take in copious amounts of water, food, energy, and consumer products; they also emit large quantities of trash and garbage. Urban areas concentrate the waste of many residents in small areas, and disposing of this waste can be challenging.

There is a long history of urban waste removal. Mesopotamian cities had wastewater management systems. The Minoans on Crete (Chapter 2) dug deep holes for garbage and covered it with dirt. The Mayans had monthly burnings of waste in villages. The Romans developed a complex and sophisticated system for removing trash, including sewers, but often simply dumped waste in vacant areas or rivers. In the 15th century, Paris implemented mandatory sweepings of streets and employed thousands of garbage collectors to take waste outside the city. Wastewater disposal in preindustrial contexts typically consisted of a privy or water closet, whose waste was carried into sewers and cesspools.

During the Industrial Revolution, trash accumulation became a severe public health hazard, encouraging the proliferation of rats and other pests and accelerating the spread of diseases such as cholera. London started the first urban solid waste management system in the late 18th century, in part to remove coal ash. Many cities were choked with horse manure; dead horses were often simply left on the street to rot. In the 19th century, it was not uncommon for urbanites to throw their trash out the window, where it accumulated on city streets, attracting dogs and feral animals. New York also dumped waste into the oceans until the early 1900s. The introduction of sewer lines built under streets and public waste removal and sanitation systems in the 19th century (starting in New York in 1895) was a major step forward in improving the lives of urbanites worldwide. By 1900, most cities had solid waste and wastewater removal systems. Early garbage removal trucks were horse-drawn, to be replaced by motorized ones in the 1930s. Landfills were introduced in the 1960s and are the primary means of waste disposal today.

Today, with large cities containing many relatively affluent people, cities produce prodigious amounts of garbage. Worldwide, urbanites generate 4 billion tons of solid waste annually. Each American generates roughly 4.5 pounds of waste daily, which for the entire country leads to 293 million tons of waste annually. New York City alone generates 12,000 tons per *day*, some of which was taken to the Fresh Kills dump on Staten Island (Fig. 7.3) before it closed in 2001. Landfills that are not properly managed generate large amounts of methane, a greenhouse gas responsible for global warming.

Typically, city garbage trucks take waste to transfer stations located in suburban areas, where it is compressed to reduce its volume. Often hazardous materials are removed at this stage to prevent leaching into groundwater later. From these stations, waste is taken to landfills, recycling plants, or incineration facilities, which can use it to generate energy. In landfills that do not receive much precipitation, waste can accumulate for years; even paper may take decades to decay. A few cities use anaerobic digesters, which degrade organic materials and generate biogas, a renewable energy source.

Wastewater is another component of urban waste, and its treatment is an important environmental and health concern. Wastewater contains a variety of suspended and chemical contaminants, pathogens, and grit. Industrial and commercial land uses generate waste that is

Figure 7.3
Dumping waste at the Fresh Kills site, New York.

different from that of households. Stormwater runoff also needs to be cleaned, especially when it picks up oils, dirt, and heavy metals from city streets. Most cities have wastewater treatment plants, which use a combination of physical, chemical, and biological processes to remove contaminants from water before it is returned to local waterways. Solid material is removed and forms sludge, which can be used for fertilizer. Oils and grease are treated separately. Secondary wastewater treatment involves biofilters and biological contactors to break down organic content. Some cities also deploy a tertiary treatment using chlorine or ultraviolet light. All of these take place under the regulatory eye of public health agencies.

The location of urban hazardous waste facilities has been the subject of much debate. Increasing evidence that such toxic centers were located in or near predominantly Black areas of many cities jumpstarted the environmental justice movement of the late 20th century. Toxic waste facilities may have been initially located in low-cost areas, which happened to be populated by people of color; subsequently their presence lowers property values. White NIMBYism (Chapter 8) was often effective at keeping them out of more well-to-do neighborhoods, whereas lower-income areas had fewer resources and access to policy makers. Housing segregation limited the ability of people of color to leave areas close to toxic waste dumps. These disparate environmental inequalities persist to the present day.

To try to minimize the output of waste, many cities have initiated recycling programs for glass, paper, metals, cardboard, and plastics. Some ask residents to separate organic from inorganic waste. Organic waste can be composted, an eco-friendly solution. Some cities have implemented waste-to-energy programs, burning waste to generate electricity, which reduces the need for landfills and minimizes methane emissions. Plastic waste is arguably the worst form, as plastic never really decomposes and can end up in the world's oceans, where it wreaks havoc with marine ecosystems. Unfortunately, only a small percentage of discarded plastic is actually recycled.

7.5 Urban Green Spaces

Although some people imagine cities as little more than forests of concrete, most cities have within them green spaces such as parks, gardens, marshes, and even forests. Green spaces are hugely important in providing shade, cooling cities, reducing energy use, dust, pollution, and enhancing urban biodiversity, and providing recreational benefits. By absorbing runoff, green spaces reduce flooding and sewer management costs by filtering wastewater. Green spaces even improve property values. Numerous studies have demonstrated that people with access to green spaces are psychologically and physically healthier than those without. Parks are common places for family outings and for people to socialize. Green spaces also provide esthetic beauty and give many urbanites their only real contact with "nature." Some have hiking and nature trails. These benefits are very real even if they are not always captured in market prices.

As cities grew exponentially during the Industrial Revolution (Chapter 3), becoming larger, denser, and dirtier, the need for green spaces grew accordingly. In New York City, Frederick Law Olmsted designed Central Park (Fig. 7.4), a large, iconic space in the middle of

Figure 7.4
Central Park in New York City is one of the most famous urban green spaces in the world. *(https://commons.wikimedia.org/wiki/File:2024–11–19-Central_Park-0669.jpg.)*

Manhattan and an important recreational site for the city. Olmstead went on to design parks in Boston, Chicago, Cleveland, Buffalo, Montreal, Washington, DC, and other cities.

Urban parks may appear to be peaceful and apolitical on the surface, but in reality they reflect the politics of urban space and who has a right to the city. In the 19th century and today, parks afforded working-class people cheap or free spaces of leisure. Park designers often envisioned them as safe spaces for White women. Few parks were built in Black areas of cities. In the New York area, the famous urban designer and administrator Robert Moses (Chapter 8) openly espoused racist views concerning parks. During the era of racial segregation in the United States, which began to decline during the Civil Rights movement of the 1950s and 1960s, African Americans were excluded from Whites-only parks. Parks, too, must thus be considered as a matter of environmental justice.

Another important form of urban green space is green belts, which typically consist of substantial areas zoned to be undeveloped or wild. These often form a ring around downtown areas (Fig. 7.5). Green belts are environmentally and socially important, providing spaces for hiking or bicycling, cooling the local atmosphere, and absorbing air pollution and storm runoff. Green belts originated in Europe, where broad boulevards separated new

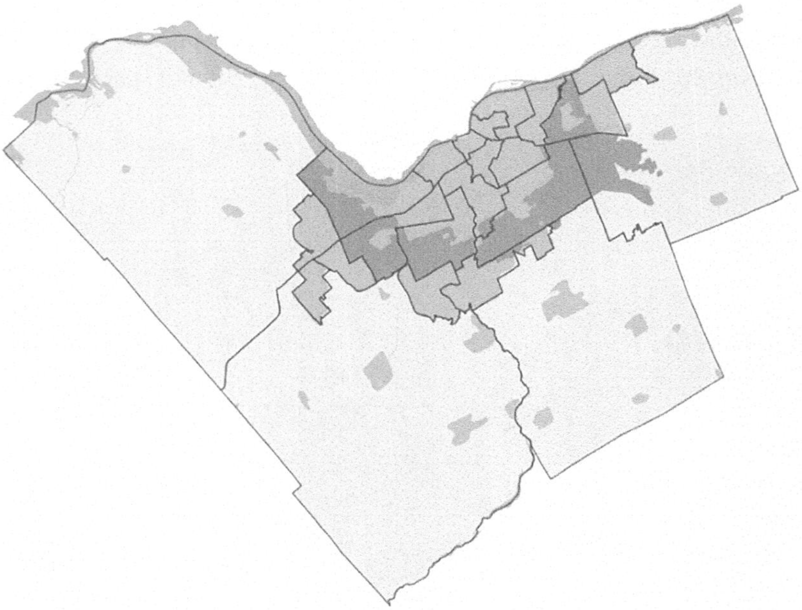

Figure 7.5

This map of Ottawa, Canada illustrates the location of an urban green belt (in green). Map lines delineate study areas and do not necessarily depict accepted national boundaries. *(https://upload.wikimedia.org/wikipedia/commons/thumb/c/c6/Ottawa_map.svg/780px-Ottawa_map.svg.png?20120123195228.)*

developments from the historic urban core. One of the most famous examples is the Ringstraße in Vienna. Often development will leapfrog over the green belt, creating satellite areas outside of the green belt.

7.6 Rural-to-Urban Land Conversion

On the edges of many cities, the transformation of agricultural land into residential and commercial uses, or rural-to-urban land conversion, is a major environmental issue. Recall the Von Thünen model of land use (Chapter 4), in which competing land uses generate the maximum rent, or profit, for a given parcel at different distances from the CBD; when the bid-rent for urban uses exceeds that of rural ones, there is a strong incentive for landowners to switch uses. As the demand for housing, retail space, roads, and suburban office parks mounts, developers eye neighboring agricultural land with ever greater interest (Fig. 7.6). As housing projects and other high-profit uses appear, farmland is swallowed up in great quantities; open fields are replaced with high density housing tracts and shopping centers. In the United States, this process destroyed about one million acres of farmland every year, although the rate has slowed recently. The precise rate of conversion reflects many factors, such as the rate of local population growth, and thus the demand for housing, the state of the local economy, the extent of transportation infrastructure in outlying areas, and zoning laws. Because many cities began as small agricultural centers, the pavement of farmland means that the best areas to grow crops—flat, well watered, with good soils—are of the first to go. In short, urban sprawl, the

Figure 7.6
Agricultural land near suburban housing development, Maryland.
(https://commons.wikimedia.org/wiki/File:Farm_land_and_housing_development,_Adamstown,_Maryland.jpg.)

dominant form of urban growth, is a major force in the destruction of farmland as well as forests and wetlands. Areas with poor or weak zoning are particularly susceptible to this process.

The conversion of farmland to urban uses entails a variety of environmental changes, which are usually assessed in Environmental Impact Statements. Drainage flows are altered, and with lower rates of evapotranspiration, more flooding is likely to ensue. Soils—a nonrenewable resource—are lost forever. The habitat for wildlife is annihilated, and biodiversity reduced. When development is discontinuous, resulting in leapfrogging sprawl near transport corridors, larger habitats are splintered into small ones, which are less amenable to the sustenance of wildlife. With less farmland, ecosystem services, or the benefits derived from healthy ecosystems such as filtering of pollution and decomposition of waste, are lessened. Moreover, the supply of farmland is also reduced, leading to a lower ability to grow crops for export or local consumption.

Socially, rural-to-urban land conversion can lead to cities swallowing up small rural villages, annihilating forms of life that may be centuries old in the process. Traditional rural architectures and cultural practices are thus sacrificed so that developers can earn a profit. On the other hand, this process also creates new supplies of housing, which is often in short supply.

For the conversion of rural land to urban purposes to occur, cities must rezone the area. Rural-to-urban land conversion can be mitigated through agricultural zoning ordinances, which are designed to protect farmland and rural amenities. Conservation projects such as parks and protected areas can limit the encroachment of cities into neighboring rural areas. Unfortunately, in most countries growth management in this regard is rather ineffective, in part because real estate developers hold considerable power at the local level. Large lot zoning, with large minimum lot sizes, accelerates low-density land uses and urban sprawl. Cities may give low, preferential property taxes to rural users. On the other hand, cities that view the conversion process positively may use eminent domain, or the state's right to acquire land even if owners do not wish to sell.

Rural-to-urban land conversion is not limited to developed countries such as the United States—it is also found in many developing countries. For example, China has experienced significant conversion on the outskirts of Beijing, Shanghai, and other cities. Developing countries with rising populations and incomes are particularly likely to witness this phenomenon.

7.7 Urban Political Ecology

Political ecology is a conceptual approach in many disciplines that takes politics seriously, attempting to overcome the false nature-culture dualism. Environmental issues are social issues, and vice versa. There can be no understanding of the causes and consequences of environmental problems without a concomitant understanding of how societies are structured, function, and change. This means that environmental problems cannot be solved through technical approaches alone, but must address social forces like economic changes, power relations, access to resources, historical context, the role of government, gender relations, inequality, and the ideological systems through which people interpret and understand the world. Traditional ecology largely ignored politics; political ecology puts it front and center. By fusing the social and the environmental, political ecology focuses on "socionature." Whereas political ecology began with the study of rural issues, it has migrated into cities. From a political ecology perspective, cities are as much a part of "nature" as are rural landscapes that appear to be pristine and untouched by people. The myth of the "natural" countryside and "unnatural" city should be discarded.

Political ecology provides useful insights into urban environmental issues such as water scarcity, air pollution, and toxic waste dumps. It examines parks and pesticide use. It has also studied issues like the political ecology of front lawns, which seems trivial on the surface but is anything but, involving labor (e.g., immigrants), pesticides, and water usage. Invariably these issues involve complex political processes, often with contingent, negotiated outcomes. Almost always, those without power—people with lower incomes or disabilities, and ethnic minority populations—tend to suffer the worst consequences.

Political ecology also called attention to the urban metabolism, the flows of materials and energy that keep the urban economy going and sustain people's lives. Cities are ecosystems in much the same way that forests, oceans, and deserts are ecosystems. Consider the material inputs needed to sustain a city: food, water, wood, paper, metals, consumer goods, petroleum, and natural gas, which are brought in by truck, rail, and airplanes. Large cities are voracious consumers of resources. These flows do not happen by themselves; they are made to happen by economic and political relations within markets and the state.

This approach to cities invites us to take into account things like energy usage and the impacts of cities on their local environments, as well as that of the world. It views urban and rural areas as inextricably locked together, so that what happens in one affects the other. Like individuals, cities have an environmental footprint that covers not only their surrounding hinterlands but, increasingly, large parts of the globe.

Taking the environment seriously also means that urban planning should prioritize sustainable development, that is, an economy and society that minimizes the use of nonrenewable resources and emphasizes renewable energy sources like wind and solar power. Sustainable development also means the widespread use of recycling to minimize waste.

7.8 The Hazardous City

Life in cities is not without its hazards. Social problems such as crime can threaten the security of everyday life. But cities also face an array of environmental hazards, such as earthquakes, hurricanes, tornadoes, and flooding. Risk is the probability that a hazard will occur; vulnerability, in contrast, is the susceptibility to harm from a risk that a hazard poses. Not surprisingly, vulnerability is closely associated with social status: lower-income and marginalized populations are always the most vulnerable. Vulnerability can be minimized by effective engineering and well-designed planning codes, disaster preparedness, and evacuation plans.

Cities located in tectonically active zones, where the planet's continental plates collide with regularity, are occasionally devasted by earthquakes. In 1906, an earthquake in San Francisco leveled the city (Chapter 3), killing 3000 and destroying 80% of its buildings (Fig. 7.7). Recent examples testify to how dangerous these can be. In 1976, Tangshan, China, suffered roughly 300,000 deaths from a major quake. Other examples include Ashgabat, Turkmenistan in 1948 (110,000 deaths), and another in Mexico City in 1985 (5000 deaths). Even minor quakes can cause extensive property damage, destroy roads and bridges, and terrify people. Earthquakes are most deadly when buildings are poorly designed and built, including the use of too much sand in the concrete, which may be the result of corruption. Earthquake mitigation plans include slope stabilization, seismic-resistant buildings, and seismic monitoring.

Hurricanes also can pose severe threats to coastal cities, causing immense property damage and loss of life within a few hours, largely due to flying debris. There is a long history of hurricanes that have hit cities on the east coast and in the southeast. In 1992,

Figure 7.7
San Francisco following the 1906 earthquake.
(https://commons.wikimedia.org/wiki/File:Sanfranciscoearthquake1906.jpg.)

Hurricane Andrew demolished 25,000 homes in southern Florida. In 2005, Hurricane Katrina devastated New Orleans, overwhelming the levees built to contain the Mississippi, killing roughly 1,400 people and causing $125 billion in damages (Fig. 7.8). Low-income neighborhoods suffered the worst. New York suffered severe flooding in the wake of Hurricane Sandy in 2012, including subway tunnels. Real estate prices in cities affected by hurricanes tend to rise significantly for several years thereafter, and hurricanes are often followed by waves of looting. Electricity, gasoline, Internet service, and drinking water supplies are disrupted for some time. Tourism to hurricane-devastated cities is muted for years. Climate change likely has intensified the frequency and severity of hurricanes. In developing countries, hurricanes and cyclones can cause catastrophic damage in urban and rural areas alike.

Several US cities have been hit by strong tornadoes, including St. Louis in 1871 and 1927; Nashville in 1933; Waco, Texas in 1953; Topeka, Kansas in 1966; and Lubbock, Texas in 1970. Early warning systems, shelters, and evacuation plans have minimized deaths due to tornadoes.

Cities are more prone to flooding than rural areas, and floods are the most common form of disaster in cities. The reason for this is straightforward, and is related to how rainfall or precipitation is distributed over portions of the Earth's surface. Essentially, precipitation can

Figure 7.8
Flooding in New Orleans in the wake of Hurricane Katrina, 2005.
(https://commons.wikimedia.org/wiki/File:Katrina-new-orleans-flooding3–2005.jpg.)

infiltrate into the ground, evaporate (or transpire through plants), or run off to another location at a lower elevation. In mathematical terms,

$$P = I + ET + R,$$

where P = precipitation

I = infiltration

ET = evapotranspiration

R = runoff

In cities, rates of infiltration tend to be low due to covered surfaces such as streets and parking lots. But the water precipitated by rainfall must go somewhere. If infiltration is reduced (and often evapotranspiration as well, due to loss of vegetative cover) by the widespread use of impermeable surfaces, then urban runoff rates must necessarily be higher. For this reason, cities tend to experience flooding more frequently and more severely than rural areas (Fig. 7.9). Not all urban floods are pluvial (caused by heavy rains); others are caused by storm surges along the coast or rivers that overflow their banks.

Figure 7.9
Urban flooding reflects the low rates of infiltration and evapotranspiration found in cities.
(https://commons.wikimedia.org/wiki/File:Urban_flooding_image.jpg.)

As cities have grown, more people have moved into flood-prone areas across the world, such as floodplains. Floods can cause catastrophic losses in lives and structural property damage. They can cause sewers to back up into toilets and basements, and lead to the proliferation of mold. Severe flooding can make streets impassable. In extreme cases, floods cause deaths due to drowning. In the United States, floods cost the country between $180 billion and $500 billion annually. Flood-prone areas face very high home insurance rates, if owners can obtain it at all. In the United States, the National Flood Insurance Program, managed by the Federal Emergency Management Agency, subsidizes insurance for homeowners in floodplains.

Cities near coasts and rivers and in low-lying areas are the most prone to flooding. Many implement mitigation measures such as levees on rivers, drainage areas, and pumps. But such strategies can be overwhelmed by storm surges, especially for cities on the coasts. Even in drier climates, flash floods brought on by intense bouts of rainfall can create havoc: Houston and Phoenix have witnessed these. And of course, urban floods are being exacerbated by rising global temperatures brought on by climate change.

Finally, epidemics and pandemics also pose grave challenges to cities from time to time. Because cities are much more densely populated than rural areas, it is easier for viruses to be transmitted among people. In the 19th century, when most cities had poor sanitation, outbreaks of cholera were common. More recently, the COVID-19 pandemic that began in 2020 took a heavy toll on cities such as New York. Public health measures can mitigate the spread of disease, such as requiring people to wear masks.

7.9 Cities and Climate Change

Anthropogenic climate change—that caused by human beings—is a severe, ongoing, worldwide crisis that threatens life as we know it. Climate change is largely attributable to the massive use of fossil fuels, which began during the Industrial Revolution and led to the emission of enormous amounts of carbon dioxide and methane, which trap solar energy. Today, cities generate roughly 70% of the gases that cause climate change. The effects of climate change are felt in rising atmospheric and sea surface temperatures, melting glaciers, rising ocean levels, and widespread disruption of ecosystems and biomes across the planet. Climate change likely also exacerbates urban heat islands, periodic heat waves, and hurricanes. It also threatens electricity and water supplies as power plants and pumping stations are put under stress.

Many cities are located near the coast and are thus threatened by rising sea levels. Even small increases in the mean sea level can lead to widespread flooding. Worldwide, more than 25 megacities are already threatened. Those located in river deltas are particularly vulnerable. In Asia, vulnerable cities include Shanghai, Guangzhou, Kolkata, Mumbai, Rangoon, Tianjin, Tokyo, Nagoya, Bangkok, and Ho Chi Minh City. Jakarta, Indonesia, is so threatened by rising tides that the Indonesian government is building a new capital city, Nusantara, in Kalimantan, Borneo. In Europe, Rotterdam and Amsterdam are particularly susceptible. In Africa, Alexandria, Abidjan, and Lagos lead the list. In North America, New York, Virginia Beach, and especially Miami are vulnerable; Miami now floods regularly whenever there is a full moon (Fig. 7.10).

Other effects of climate change include more frequent and severe storm surges, flooding, and heat waves. It can cause saltwater intrusion into freshwater groundwater supplies. Rising temperatures affect electrical and water supply systems as demand rises, leading to shortages. It can exacerbate the spread of diseases, especially those spread by mosquitoes and polluted water supplies. More extreme weather events will lead to greater damage from high wind speeds. Everywhere, the poor will be the hardest hit by these changes. In the developing world, climate change particularly threatens those living in slums and shantytowns.

There is a significant global movement to acknowledge and plan for climate change. As national governments are often reticent to do anything about this existential crisis, leadership in combating climate change has fallen to urban areas. Wealthier and better educated cities tend to be at the forefront of this movement. Cities around the world are engaged with climate change policy networks and taking steps such as reducing carbon emissions, building green infrastructure and reflective roofs, planting trees, and encouraging renewable sources of energy. Many cities encourage the use of solar panels on roofs, which provide low-cost and pollution-free sources of carbon-free energy and also reduce noise pollution. Even simply

Figure 7.10
Downtown Miami tidal flooding.
(https://commons.wikimedia.org/wiki/File:Miami_proper_tidal_flooding_11.jpg.)

painting roofs white can cool a city a few degrees. Climate change adaptation strategies have become increasingly common among urban planners, who encourage higher density land uses near public transport routes. Some have imposed carbon taxes to reduce emissions. Other steps include encouraging green roofs with vegetation on them, which reduce temperatures and air pollution. Often known as resilience, adapting and planning for climate change will be a worldwide imperative in the future. Networks of cities allow them to share expertise and learn from each other's experiences.

7.10 Chapter Summary

Cities are always and everywhere embedded in local, regional, and global biophysical environments, or what we commonly call "nature." The biophysical world has been repeatedly and extensively altered by human beings since they first emerged. Although cities may at first glance appear to be the antithesis of "nature," in reality they are shaped by—and in turn shape—the environments around them.

An important facet of the environmental side of cities is their local climates. With land surfaces that are generally paved, cities reflect much more energy from sunlight into the air than do rural areas. The result is that cities are typically several degrees warmer than surrounding areas. This notion is captured in the idea of urban heat islands.

Cities have suffered air and water pollution essentially since they first began. Particularly with the Industrial Revolution in the 19th century, the air quality of cities contained large amounts of particulate matter from smoke, and urbanites' health conditions suffered accordingly. The widespread use of the automobile added several noxious gases to urban atmospheres. Although most countries have taken steps to reduce air pollution, hundreds of millions of people worldwide continue to suffer respiratory problems as a result. The quality of air is typically worst in areas populated by underrepresented communities and those with lower incomes, making air pollution a matter of environmental justice.

Large cities need vast quantities of water. The urban modernization projects that began in the 19th and early 20th centuries included a large number of reservoirs and aqueducts to ensure a steady supply of water. Cities in places where water is scarce encourage residents to save water by minimizing its use in watering lawns, swimming pools, and car washes. In developing world cities, obtaining water is a difficult task for one-quarter of the world's population.

With numerous people, particularly in wealthy societies in the developed world where mass consumption reigns supreme, cities produce vast quantities of waste and garbage. In the 19th century, garbage on the streets became a severe public health problem. Managing waste quickly and safely is an important task. A considerable body of research has demonstrated that communities with lower incomes and minority ethnic populations are far more likely to be exposed to toxic waste, with detrimental health effects, making it a matter of environmental justice. Today many cities reduce their volume of waste by encouraging recycling, although at times the effectiveness of this approach is questionable.

Green spaces in cities, such as parks, provide important places for health and recreation. Trees cool surrounding areas, provide shade, and clean the air. Historically, access to parks was differentiated by class and race, making this a matter of environmental justice as well. Wealthier areas usually have more trees than do lower-income ones. Some cities plan and design green belts on the urban periphery.

On the outskirts of cities across the world, rural land is being converted to residential and commercial purposes at rapid rates. Rural-to-urban land conversion results in the annihilation of farmland, the loss of rich agricultural soils, and reduced biodiversity. Poor planning leaves this process uncontrolled; other cities attempt to reduce the rate through agricultural zoning ordinances.

An important way of examining urban environmental issues is urban political ecology, which views questions of land use and land cover change, access to services, and inequalities regarding waste, water, and clean air through the lens of social theory. In this light, environmental issues are always social and political ones, and vice versa. This perspective sheds light on the power dynamics that always and inevitably underpin urban growth, decline, and the ways in which urban space is structured.

Cities are frequently visited by hazards such as earthquakes, hurricanes, and floods. The underresourced and underrepresented always suffer the most. Because they contain large amounts of impermeable land, runoff in cities is higher than in rural areas, and consequently so are floods, which regularly create significant loss of life and property damage, particularly in coastal areas. "Natural" hazards are not so natural after all. Climate change has increased the severity of hurricanes, and may play a role in urban flooding as well.

Since the Industrial Revolution began, humans have relied on fossil fuels, emitting vast quantities of gases such as carbon dioxide and methane, which trap solar energy, causing anthropogenic climate change. The world has grown noticeably warmer in the last century, and continues to do so. Cities around the world are threatened by climate change, particularly those in low-lying coastal areas. Many megacities worldwide are threatened by rising sea levels. Climate change also plays a role in the spread of diseases and urban heat islands, and rising temperatures pose severe challenges for electrical grids and water supply systems.

Study Questions

1. What is an urban heat island and what causes it?
2. Why is air pollution a matter of environmental justice?
3. What measures can cities take to mitigate severe water shortages?

4. How are toxic waste dumps related to the social geography of the city?
5. Why are green spaces important? Is the distribution of trees related to class?
6. What drives rural-to-urban land conversion?
7. Summarize the perspective of urban political ecology.
8. How do hazards differentially affect people of different ethnicities and class?
9. What are some of the ways that climate change is endangering cities?

Further Reading

S. Bell, Urban water sustainability: Constructing infrastructure for cities and nature Urban Water Sustainability: Constructing Infrastructure for Cities and Nature. Taylor and Francis, United Kingdom Taylor and Francis, United Kingdom (2017), 1–185, Available from: http://www.tandfebooks.com/doi/book/10.4324/9781315680811. https://doi.org/10.4324/9781315680811.

Christopher G. Boone, Geoffrey L. Buckley, J. Morgan Grove, Chona Sister, Parks and people: An environmental justice inquiry in Baltimore, Maryland. Annals of the Association of American Geographers. 0004–5608. 99 (4) (2009), 767–787, Available from: https://doi.org/10.1080/00045600903102949.

H. Bulkeley, Cities and climate change Cities and Climate Change. Taylor and Francis, United Kingdom Taylor and Francis, United Kingdom (2013), 1–268, Available from: http://www.taylorandfrancis.com/books/details/9780203077207/. https://doi.org/10.4324/9780203077207.

Steven J. Burian, Stephan J. Nix, Robert E. Pitt, S. Rocky Durrans, Urban wastewater management in the United States: Past, present, and future. Journal of Urban Technology. 1063–0732. 7 (3) (2010), 33–62, Available from: https://doi.org/10.1080/713684134.

Joel E. Cohen, Cities and climate change: A review essay. Population and Development Review. 0098–7921. 45 (2) (2019), 425–435, Available from: https://doi.org/10.1111/padr.12259.

L.K. Comfort, Cities at risk: Hurricane Katrina and the drowning of New Orleans. Urban Affairs Review. 15528332. 41 (4) (2006), 501–516, Available from: https://doi.org/10.1177/1078087405284881.

M. Davis, *Ecology of Fear: Los Angeles and the Imagination of Disaster*. Verso (2022).

Kelsey N. Ellis, Linda M. Sylvester, Jill C. Trepanier, Spatiotemporal patterns of extreme hurricanes impacting US coastal cities. Natural Hazards. 0921–030X. 75 (3) (2015), 2733–2749, Available from: https://doi.org/10.1007/s11069–014–1461–4.

M. Ernwein, Bringing urban parks to life: The more-than-human politics of urban ecological work. Taylor and Francis Ltd., United Kingdom Annals of the American Association of Geographers. 24694460. 111 (2) (2020), 559–576, Available from: http://www.tandfonline.com/loi/raag21#.VyMsnk1f3cs. https://doi.org/10.1080/24694452.2020.1773230.

Volker Heidt, Marco Neef, Benefits of Urban Green Space for Improving Urban Climate . Springer Science and Business Media LLC, Springer Science and Business Media LLC, 84–96, https://doi.org/10.1007/978–0–387–71425-7_6.

A. Hunt, P. Watkiss, Climate change impacts and adaptation in cities: A review of the literature . Kluwer Academic Publishers, United Kingdom Climatic Change. 01650009. 104 (1) (2011), 13–49, Available from: http://www.wkap.nl/journalhome.htm/0165–0009. https://doi.org/10.1007/s10584–010–9975–6.

Jinwon Kim, Changwook Kim, Seungji Lee, Ji Youn Jeong, Race, poverty, and space: A spatial intersectional approach to equity of urban park access. Cities. 02642751. 147 (2024), 104819, Available from: https://doi.org/10.1016/j.cities.2024.104819.

I. Leahy, Y. Serkez, Since when have trees existed only for rich Americans?. (2021), Available from: https://www.nytimes.com/interactive/2021/06/30/opinion/environmental-inequity-trees-critical-infrastructure.html.

R.I. McDonald, K. Weber, J. Padowski, M. Flörke, C. Schneider, P.A. Green, T. Gleeson, S. Eckman, B. Lehner, D. Balk, T. Boucher, G. Grill, M. Montgomery, Water on an urban planet: Urbanization and the reach of urban water infrastructure . Elsevier Ltd, United States Global Environmental Change. 09593780. 27 (1)

(2014), 96–105, Available from: http://www.elsevier.com/inca/publications/store/3/0/4/2/5. https://doi.org/10.1016/j.gloenvcha.2014.04.022.

P. Mohai, R. Saha, Racial inequality in the distribution of hazardous waste: A national-level reassessment. Social Problems. 15338533. 54 (3) (2007), 343–370, Available from: https://doi.org/10.1525/sp.2007.54.3.343.

C. Philo, Animals, geography, and the city: notes on inclusions and exclusions. Environment & Planning D: Society & Space. 13 (6) (1995), 655–681.

Laura Pulido, Rethinking environmental racism: White privilege and urban development in Southern California. Informa UK Limited, Informa UK Limited (2017), 379–407, Available from: https://doi.org/10.4324/9781315256351-17.

Anne Rademacher, Urban political ecology. Annual Review of Anthropology. 0084–6570. 44 (1) (2015), 137–152, Available from: https://doi.org/10.1146/annurev-anthro-102214-014208.

Ahmed Memon Rizwan, Leung Y.C. Dennis, Chunho Liu, A review on the generation, determination and mitigation of Urban Heat Island. Journal of Environmental Sciences. 10010742. 20 (1) (2008), 120–128, Available from: https://doi.org/10.1016/s1001-0742(08)60019-4.

Kirsten Schwarz, Michail Fragkias, Christopher G. Boone, Weiqi Zhou, Melissa McHale, J. Morgan Grove, Jarlath O'Neil-Dunne, Joseph P. McFadden, Geoffrey L. Buckley, Dan Childers, Laura Ogden, Stephanie Pincetl, Diane Pataki, Ali Whitmer, Mary L. Cadenasso, Steven Arthur Loiselle, Trees grow on money: Urban tree canopy cover and environmental justice. PLOS ONE. 1932–6203. 10 (4) (2015), e0122051, Available from: https://doi.org/10.1371/journal.pone.0122051.

E. Swyngedouw, *Social Power and the Urbanization of Water: Flows of Power*. Oxford University Press (2004).

E. Swyngedouw, The political economy and political ecology of the hydro-social cycle. Journal of Contemporary Water Research & Education. 1936–7031. 142 (1) (2009), 56–60, Available from: https://doi.org/10.1111/j.1936-704x.2009.00054.x.

E.M. White, A.T. Morzillo, R.J. Alig, Past and projected rural land conversion in the US at state, regional, and national levels . Elsevier, United States Landscape and Urban Planning. 01692046. 89 (1–2) (2009), 37–48, Available from: http://www.elsevier.com/inca/publications/store/5/0/3/3/4/7. https://doi.org/10.1016/j.landurbplan.2008.09.004.

Y. Zhang, Z. Yang, X. Yu, Urban metabolism: A review of current knowledge and directions for future study. American Chemical Society, China Environmental Science and Technology. 49 (19) (2015), 11247–11263. 15205851. Available from: https://doi.org/10.1021/acs.est.5b03060, http://pubs.acs.org/journal/esthag.

CHAPTER 8

The State and Urban Structure

Chapter Outline

Cities under capitalism are the products of both markets and the state, or government in all its forms. Despite popular but erroneous misunderstandings that capitalism consists solely of a "free market," the reality is that capitalist societies rely heavily on the state for many things. The state does what the market cannot: it addresses market failures; it guarantees property rights; it provides public goods and services; it sets forth laws and regulations; and it acts on the international stage, shaping trade, immigration, and currency exchange rates. The capitalist city, in short, would be impossible without the state. The nature and degree of state intervention have changed over time, occasionally growing during periods of crisis (e.g., the Great Depression of the 1930s) and sometimes retreating, as during the current neoliberal era of deregulation.

This chapter examines how the state shapes cities in several ways. It begins with an overview of national urban policy, or how the federal government affects cities in several ways. Second, it examines urban planning, in which municipal governments shape urban land use, economies, and daily life. The third part focuses on zoning, the legal restrictions over urban land use. Fourth, it summarizes the role of the state in housing markets.

8.1 National Urban Policy

National governments have enormous impacts on cities through a variety of channels. These include programs such as housing, poverty alleviation, environmental regulations, and attempts to reduce regional disparities and encourage economic development. Increasingly, many governments around the world have begun to address climate change and its impacts, including sustainable development, particularly in light of the United Nations' Sustainable Development Goals. The Organization for Economic Cooperation and Development, a global network to encourage trade and prosperity, outlines several criteria for national urban policies,

Urban Geography. DOI: https://doi.org/10.1016/B978-0-443-30281-7.00008-6

such as maximizing cities' ability to achieve their potential, protect the environment, and provide adequate housing. Most countries have national urban plans, which involve long-term planning for land use, housing, transportation, and air and water quality. However, the extent and nature of national policies vary widely among countries. In contrast to the United States, European countries tend to have well-developed national urban policies, such as strict controls over land use.

In a sense, *any* national policy (e.g., population, crime, transportation) will affect cities more than most places, often simply because most people live in cities. Some national initiatives have enormous urban impacts even if they are not explicitly urban in their focus. For example, in the United States, the Federal Interstate Highway System, which started in the 1950s and was completed in the mid-1970s, had major impacts on cities, including facilitating interurban trade and migration and encouraging suburbanization (Fig. 8.1). By making it much easier for people and goods to move among cities, the highway system facilitated interurban interaction and helped cities to realize their local comparative advantages and export bases (Chapter 4).

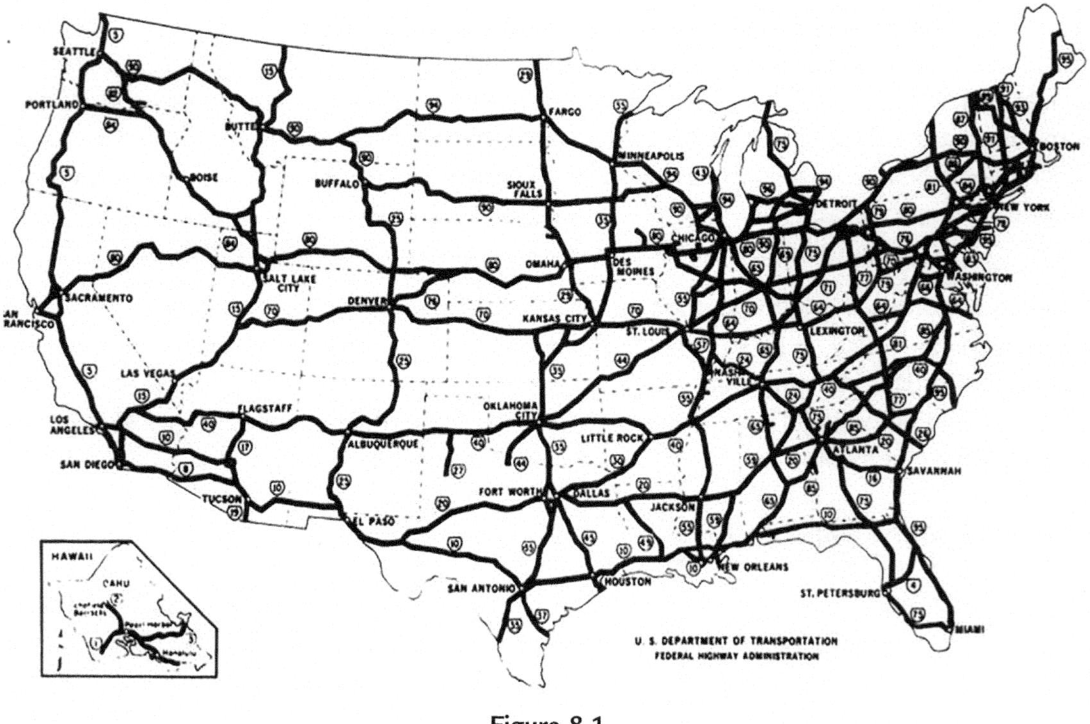

Figure 8.1
The Federal Interstate Highway System. Map lines delineate study areas and do not necessarily depict accepted national boundaries.
(https://commons.wikimedia.org/wiki/File:Interstate_Highway_System_Map.jpg.)

Ironically, in many countries, national policies are heavily directed toward rural areas, largely to encourage development there, including subsidies to farmers. However, the dynamics of capitalism have tended to favor cities over surrounding areas (Chapter 4). Thus, many governments have programs that actively try to compensate for urban-rural differences. For example, they may subsidize farmers with crop prices and insurance; offer rural poverty alleviation programs; and encourage telecommunications companies to lay fiber optic cables there. In the United States, rural states tend to receive far more from the federal government than the government receives in taxes. Roughly 15% of the population lives in rural areas, but two-thirds of all road miles are located there.

The federal government has been active in housing policy, which greatly shapes livability in urban areas. Many of these arose as part of President Roosevelt's New Deal during the Depression of the 1930s. In 1932, the government created the Federal Home Loan Bank to subsidize mortgages for lenders. In 1934, the National Housing Act created the Federal Housing Administration and allowed for the interest on mortgages to be tax-deductible. In 1938, it created the Federal National Mortgage Association (FNMA), commonly known as Fannie Mae, one of the largest companies in the world, to expand the mortgage market by issuing mortgage-backed securities. Similarly, in 1970 it founded the Federal Home Loan Mortgage Corporation (FHLMC), commonly known as Freddie Mac, to buy mortgages, pool them, and sell them as mortgage-backed securities. In the 1950s the federal government initiated urban renewal programs (see below) that cleared many low-income people from inner cities to make them ripe for redevelopment. In 1965 the Housing and Urban Development Act subsidized housing and led to the growth of public housing. In the 1960s, President Johnson's Great Society programs had many urban effects. The so-called "war on poverty" saw many programs, such as Job Corps, to improve life in the ghettos.

Federal urban policy underwent a significant change in the 1980s as the Reagan Administration implemented a radically different set of policies. The federal government's role in local urban affairs declined and that of local governments increased. Funding for federal housing projects declined rapidly. One step was the initiation of Community Development Block Grants to distressed cities, aimed at fighting poverty, increasing affordable housing, and infrastructural development. These gave considerable leeway to localities in how to use their funds.

Another federal program involved Enterprise Zones, designed to unleash market forces in cities by removing government land use controls, labor training programs, and tax incentives (Fig. 8.2). Enterprise zones became very popular throughout the United States and parts of Europe, but never lived up to their promise of revitalizing inner cities. Critics denounced them as gifts to the business community that lacked the priority of helping underserved populations. Some held that enterprise zones were simply a means of subsidizing gentrification.

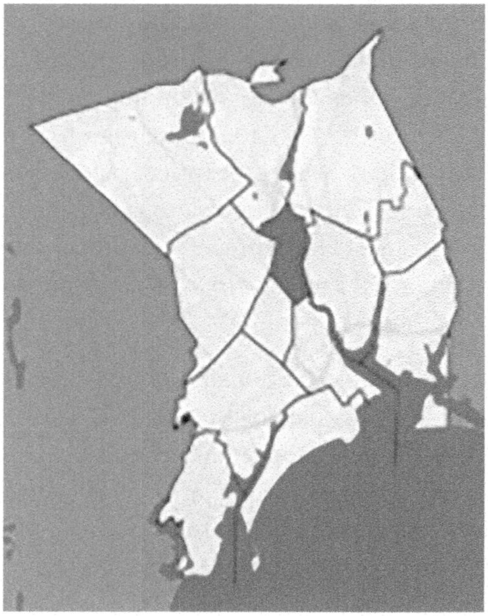

Figure 8.2
Map of the enterprise zone in Bridgeport, Connecticut. Map lines delineate study areas and do not necessarily depict accepted national boundaries.
(https://commons.wikimedia.org/wiki/File:Bridgeport_neighborhood_map_-_Enterprise_Zone.png.)

8.2 Urban Planning

One of the most important ways in which the state is involved in shaping urban space is urban planning (also called town, city, or regional planning). Planning involves many insights from geography, engineering, architecture, public administration, and urban design. It is a social as well as technical activity.

Planning is invisible to most people; it occurs behind the scenes. Yet it has enormous implications for how well cities function, shaping the everyday lives of countless urbanites. Planners allocate public resources and fix problems that the market cannot (e.g., negative externalities, which are market failures). Examples of market failures include inefficient traffic patterns and the destruction of the environment, activities for which the cost is not reflected in the market price. Planners regulate urban land use patterns; oversee the design of roads, highways, water lines, and the electrical grid; design and implement zoning ordinances; and affect the location and nature of urban public services like hospitals, schools, and bus routes. The goal is to make urban areas more livable, enhance the quality of life, and reduce the problems that markets often create. Left to their own devices, markets often produce anarchical or inefficient patterns of land use.

Urban planning has a long history. Some classical Greek and Roman cities were planned. Medieval cities were planned, with a focus on the central plaza. In the 19th century, European cities were planned to accommodate the requirements of industrial capitalism, including a modern infrastructure. As the Industrial Revolution accelerated urban growth, a new type of city was born. A spectacular example of this process was the redesign of Paris by Georges-Eugène Haussmann in the 1850s. Chosen by Emperor Napoleon III, Haussmann oversaw a massive public works effort that cleared away most of the city's medieval neighborhoods and put in wide boulevards that enhanced the mobility of goods and people, improved sanitation, traffic circulation, railroad stations, water pipes, waste removal, parks, streetlights, and an electrical grid. As a result, Paris became known as the "City of Lights," the embodiment of modernity. Haussmann's efforts are still visible in the Parisian landscapes of today (Fig. 8.3).

Early planning in the United States included L'Enfant's design of Washington, DC (Chapter 3). However, planning in the United States unfolded in a very different cultural context than it did in Europe. American society tends to be much more conservative, individualistic, and it exhibits a sustained faith in markets and distrust of the state more so than do Europeans. As a result, planning in the United States developed later, and tends to be much more local in nature, whereas many European countries have national planning agencies. The Progressive Era of the late 19th

Figure 8.3
An aerial view of Paris, showing the broad boulevards constructed by Haussmann in the mid-19th century.
(https://commons.wikimedia.org/wiki/File:Aerial_photograph_of_Paris_(14619557430).jpg.)

and early 20th centuries gave a great impetus to planning by enabling it to confront the ills of industrial cities, including sanitation and a growing infrastructure. The Chicago World's Fair of 1893 was a major step in this regard. By the 20th century, planning had become professionalized. For example, the government's right to regulate privately owned land had been legally established, and zoning began in New York City in 1916.

Planners conduct a variety of tasks that serve the public in different ways. They monitor the city's environment, including its population and economy, and use forecasts of the future. This may involve making maps and using geographical information systems, and tracking things like property tax revenues and outbreaks of diseases. Planning is fundamentally concerned with change and anticipating what is likely to happen in the near future. Planners may run simulations and models, and may conduct economic impact studies about the repercussions of various events, such as large firms opening or shutting down.

Transportation planners keep track of the quality of road surfaces and the volumes of traffic, seeking to reduce delays, minimize congestion, lower air pollution, and match the supply and demand for parking spaces. They lay out the routes for biking paths and pedestrian walkways. They are well aware that transportation and land use systems deeply shape one another, and change over time. Some planners are involved in disaster preparedness, including evacuation plans in cases of emergency (e.g., major earthquakes). Planners help to optimize the delivery systems of public services, including the locations of schools, parks, water mains, recycling facilities, waste incineration plants and disposal sites, and homeless shelters. They are vital to helping cities understand how many facilities they need and where they should be located. They may meet the public to solicit their input about issues such as school closures or the locations of new facilities, often encountering the NIMBY syndrome (Chapter 6).

Some planners are involved in economic development, learning what firms need, preventing others from leaving, and considering how the local labor force changes. They may be involved in tourism promotion, downtown redevelopment, or obtaining state and federal government funds. They oversee growth management, ensuring that the infrastructure and land use changes are in balance. Other planners are involved in environmental management, preventing the loss of wetlands and endangered species, as well as flood mitigation. They often try to protect agricultural land from rural-to-urban land conversion. Some planners focus on green spaces, including parks, hiking trails, and nature preserves. Bad planning leads to haphazard development, such as the appearance of countless strip malls, and degrades the environment.

Environmental planning has surprisingly deep roots. At the start of the 20th century, Ebenezer Howard, a British planner, started the Garden City movement, which envisioned cities well provisioned with green spaces as an antidote to the ills of the Industrial Revolution. He envisioned a series of parks and gardens strategically located throughout the city (Fig. 8.4). Around the same time, in the United States the City Beautiful movement advocated for public

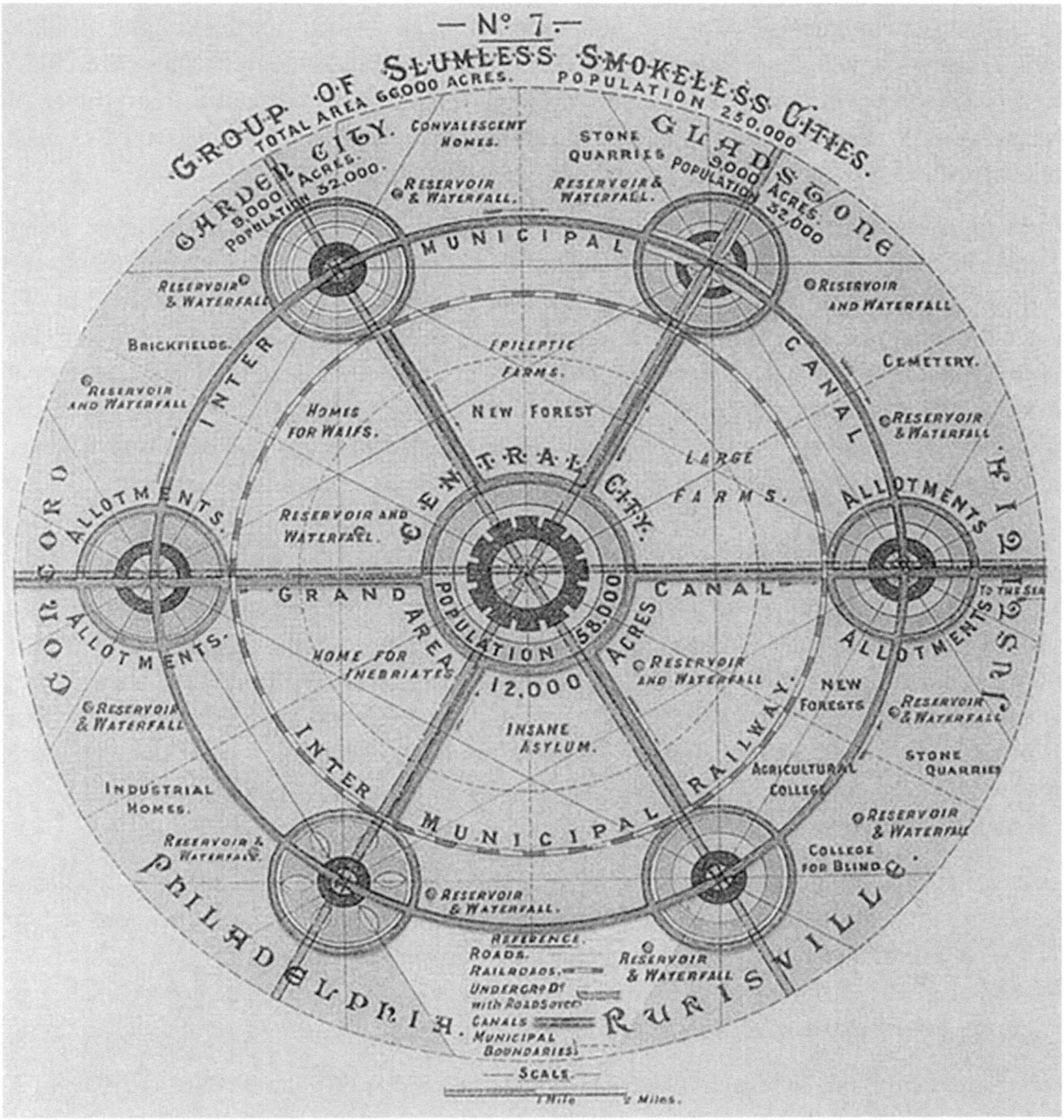

Figure 8.4
Ebenezer Howard's vision of the garden city.
(https://commons.wikimedia.org/wiki/File:Garden_City_Concept_by_Howard.jpg.)

housing and monuments, not only to improve the urban environment but also, allegedly, to create moral virtue among urbanites. Environmental planning flourished after the birth of the modern environmental movement in the 1960s, and became concerned with a wide variety of issues such as agricultural land, green spaces, renewable energy, recycling, and habitat protection. The adoption of Environmental Impact Statements, which detail the economic, social, and environmental costs of new projects, was critical in this regard. Today, environmental planning is largely concerned with sustainability, helping to minimize cities' impacts on the environment, and promoting long-term growth. Of course, the specter of climate change hangs over this issue everywhere (Chapter 7). Planners in coastal cities must contend with rising sea levels and the flooding that they entail.

Urban planners have worked under different paradigms or models at various times. These involve differing notions of the role of planners and the public. By far the most common is the rational comprehensive model, widely regarded as the orthodox or mainstream view. In this perspective, the planner is an expert on urban problems who uses their expertise to help the public. Planners are, or at least should be, value-free and objective, never letting their personal views enter into planning decisions. Planning is said to serve the public interest, or do the most good for the greatest number of people. There is, of course, the question as to whether there is only one public interest or if the public even knows what is in its interest.

The rational comprehensive approach begins by soliciting residents' views of their environment, listening to their priorities and the problems they want to see addressed. The planner helps the public to clarify its goals, and then suggests alternative plans that would reach them. Planners then adopt a single, comprehensive plan—usually spelled out in a thick volume—and strategies to implement it. The plan is usually limited in its time horizon to a decade or two. Planners then open the plan to public hearings and periodic review. The best planning works from the bottom up, not the top down, and involves decision-making by those most affected. The rational comprehensive model favors a cautious, incremental, step-by-step approach to planning.

While it is broadly popular, the rational comprehensive approach to planning also has weaknesses. For example, who gets to define "the" public interest? Is there only one public interest or several incompatible ones? Homeowners, developers, environmentalists, and others have dramatically different priorities. The view suggests that municipal governments can anticipate problems when in fact they usually react to them after the fact.

More importantly, this approach is naïve regarding the matter of politics. Planners may wish to stay out of politics, but planning and politics are deeply intertwined. In the United States, where most decisions are made and enforced at the local level, planning is fragmented among tens of thousands of counties and municipalities (in Europe, national planning is more common). Planners work within the executive branch of government; they are not elected and therefore are not legally empowered to enforce decisions. Planners must work with politicians, who may have their own priorities. For example, planners often prioritize the long run, whereas politicians prioritize getting reelected in the short run. Planners favor clear goals, objectives, and tradeoffs; politicians often want vague goals to avoid alienating potential supporters.

8.3 Urban Public Services

A very important way in which the state shapes urban space is through the provision of public goods and services, such as roads, police, firefighting, trash removal, snow plowing, libraries, jails, parks, homeless shelters, public schools, public hospitals, public swimming pools, and public transportation. Clearly there is often overlap and competition between some public and private services, such as health care, transportation, and education. Public services make life in the city possible, enabling movement, health, recreation, and education.

The economics of public goods and services differ from those of private ones. Public goods are usually funded by bonds (usually at the state or county level), while public services are usually funded by property taxes. Many public goods are funded by the federal government, such as the interstate highway system, as are some services (e.g., defense, federal prisons, some large hydroelectric dams; and the post office, which is a quasipublic institution). Public services at the federal level should not be confused with transfer payments, such as Social Security or veterans' benefits. Some public goods are funded by state governments, such as state highways and state prisons. Public goods and services are socially necessary, but often not profitable (there are a few exceptions, such as privately run roads), so they are built, delivered, and maintained by the state. Most public services in the United States are funded at, and provided by, county and municipal governments.

For privately owned commodities, there is a legal right to exclusion. A homeowner can rightfully refuse to let someone enter their property. For public goods and services, which are paid by either taxpayers or bond holders, this is not the case. One person, for example, cannot legally keep another off of the sidewalk or freeway.

By far the most important and expensive public service is education. In the United States, without a national school system, schools tend to be funded mostly through local property taxes, which results in wide variations among school districts in the amount of funds available, and thus the quality of education (Fig. 8.5). While variations in funding per student are small within the same city, they tend to be large among cities. Well-funded districts tend to have smaller classes, relatively well-paid teachers, and good facilities, while the reverse is true in poorer ones. This process has produced large discrepancies between the public schools in inner cities, whose governments are strapped for funds and must allocate monies to other services (e.g., subways and buses), and suburbs, which tend to be wealthier and therefore have better schools. It has also resulted in severe racial inequalities in funding, as children of color tend to live in lower-income communities. Because schools are one of the largest motors of social mobility, inequalities in the quality of education among school districts and cities tend to reinforce rather than overcome socioeconomic inequalities.

Public hospitals are another important public good. Health care is an enormous part of the American economy, comprising 20% of the Gross Domestic Product. Increases in health care

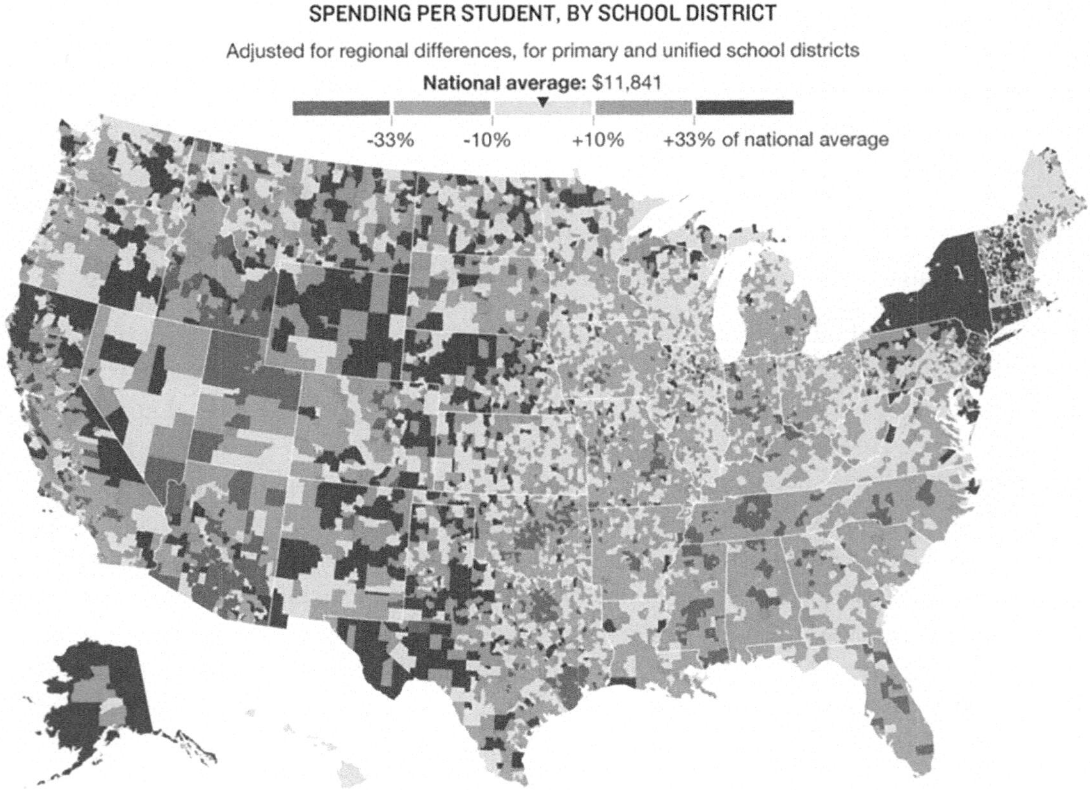

Figure 8.5
Funding per student by school district compared to the national average.
(https://commons.wikimedia.org/wiki/File:Funding_of_K-12_schools_map.png.)

are driven by several factors, including an aging population. Concerns over the number of people without health insurance have loomed large in national policy debates for decades. Public hospitals, which compete with privately owned ones (both for-profit and nonprofit), are critical places to deliver health care services to low-income populations. Many people who lack health insurance or are poorly insured use emergency rooms as their primary health care facilities; these are often overcrowded, with long waiting times.

Another important public service is transportation, including buses and subways. As seen earlier (Chapter 3), there is a long history of urban transportation. In the 19th century, privately owned omnibuses were common. By the late 19th century, as electrical lines were installed, privately owned light rail systems enabled early suburbanization. But publicly owned mass transit largely began in the 1890s when the largest cities undertook construction of light rail systems. Many of these were sold or driven out of business in the 1920s with the shift to the automobile.

Large cities such as New York, Boston, Philadelphia, and Chicago began building subway systems in the early 20th century. Europeans erected them even earlier. Even notoriously auto-dependent Los Angeles has a small subway system today. Subways are very expensive to build and maintain, and therefore require large volumes of passengers to be financially feasible. For this reason they are always found in large metropolitan areas. Even then, they require subsidies to keep operating. The New York subway system, which started in 1904 and is the largest in the world, handles two billion rides per year (Fig. 8.6). In Washington, DC, the metro system was built and operated by the federal government.

Buses are the most common form of public transportation services. Their roots extend to the mid-19th century, when horse-drawn omnibuses provided transportation to workers in growing industrial cities (Chapter 3). Buses are well suited to cities because they are relatively cheap and flexible; their routes can be changed as needed. Different types of bus service serve different riders, including transit buses, park and ride, feeder buses that drop passengers off at transit points, and rapid commuter express buses from the suburbs. There are also school buses and intercity bus services.

A fundamental constraint to public transportation is the competition from privately owned cars. Roughly 90% of all commutes in the United States are by car, and most car trips involve a single person. In contrast, mass transit accounts for only 6% of all commutes, and these are highly concentrated in a few large cities.

Why are cars so popular? Other than convenience, the short answer is that cars are much quicker than mass transit systems. The biggest cost to riders is the time they spend commuting, and because cars save time, they save money. To illustrate this notion, consider the following hypothetical example. A worker earns $24 per hour. Say the out-of-pocket expenses for cars and buses are the same, for instance, $5. But a car trip that takes 10 minutes costs the worker $6, whereas a bus trip is much longer, costing $12. When the two sets of costs are added up, cars are always cheaper. The cost savings of car versus bus trips become even greater as incomes rise. For this reason, mass transit in most cities is mostly used by lower-income people. (In New York, where driving is very expensive and difficult, the middle class uses mass transit regularly). When incomes rise, riders will shift into cars because the time savings are so large. This means that in order to make mass transit more attractive, cities should make buses quicker, not cheaper.

Example: worker earning $24/hour:

	car	bus
Out of pocket costs	$5	$5
Time	10 min $6	30 min $12
Total	$11	$17

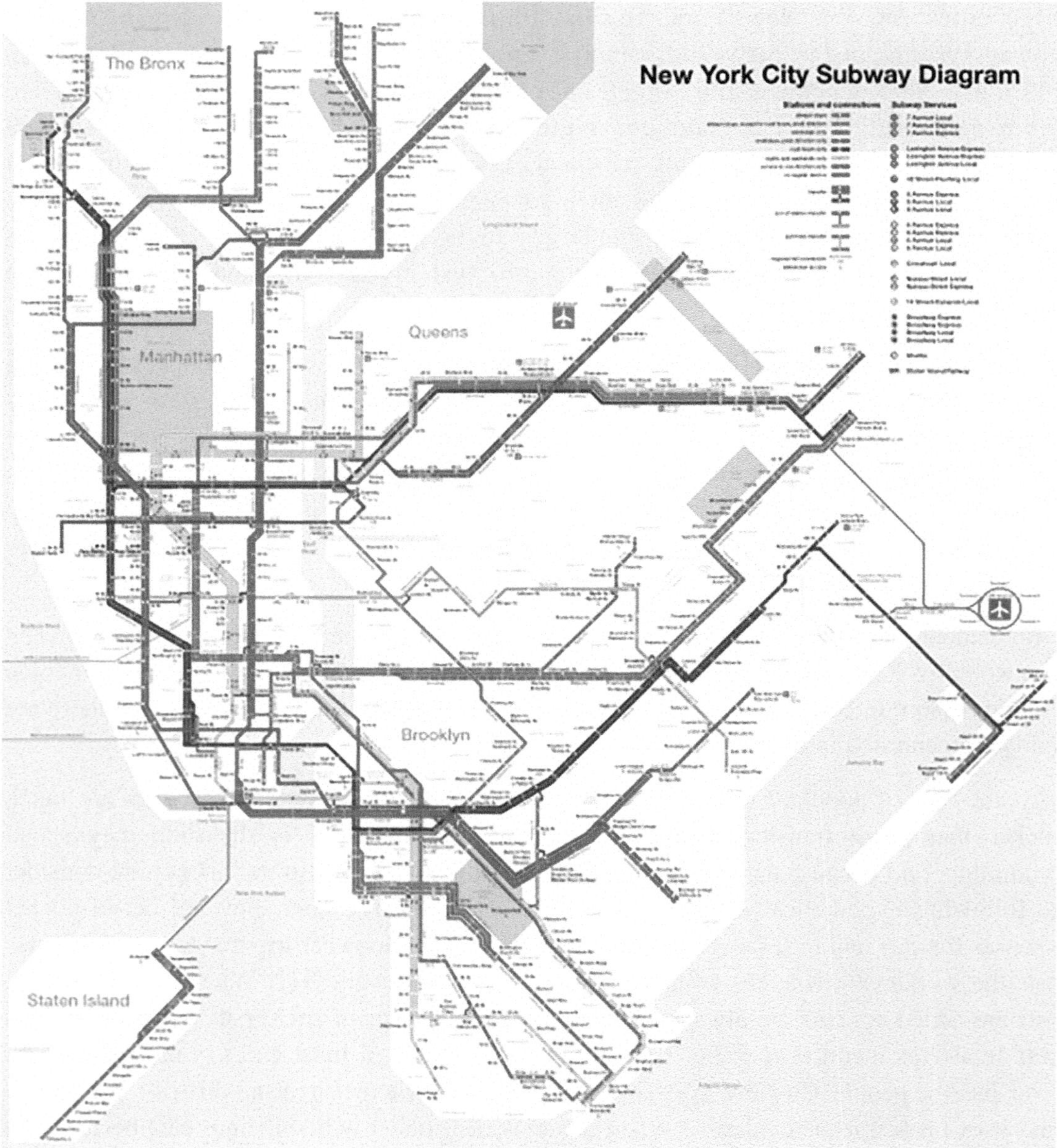

Figure 8.6
Map of the New York Subway System. Map lines delineate study areas and do not necessarily depict accepted national boundaries.
(https://commons.wikimedia.org/wiki/File:NYC_subway-4D.svg.)

All public transport services lose revenue, that is, the income from tickets never covers the operating costs. The reason for this is that demand for transportation is highly uneven throughout the day. Demand is maximized during the morning and evening rush hours, when

Figure 8.7
A public bus in Laredo, Texas is mostly empty except during rush hour traffic.
(https://commons.wikimedia.org/wiki/File:Laredo_bus_interior.jpg.)

subways and buses are filled to capacity. Other than these times, however, they often are nearly empty (Fig. 8.7). Most of the time, mass transit systems are underutilized. However, even when these systems have few riders, cities must still pay drivers, gasoline, debt service, and insurance. Financially, therefore, public transport services are a black hole. For example, the New York Metropolitan Transit Authority loses between $2.6 billion and $3 billion per year; tolls and fares only cover 35% of its operating expenses, and the difference is paid by state, municipal, and federal government subsidies.

The reliance of mass transit on high volumes of passengers to cover its operating income raises the question of when bus and subway systems become more cost-effective than cars. Fig. 8.8 portrays the cost per rider for the three major modes. For cars, the cost is relatively invariant with respect to the volume of trips. For buses, costs are high but decline as ridership increases, breaking even with cars around 10,000 passengers per hour. And subways, which

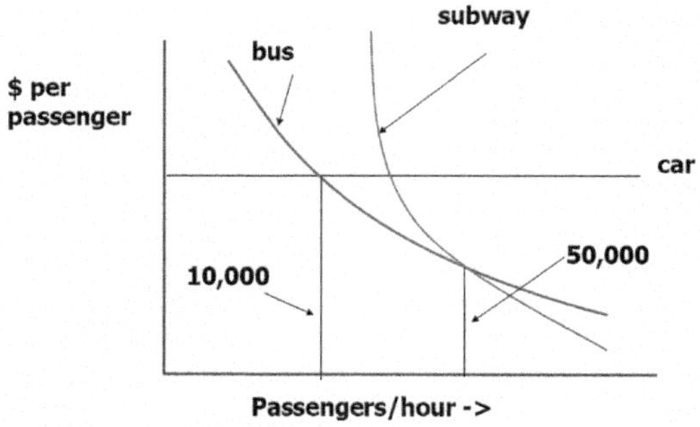

Figure 8.8
Economics of cars versus bus and subway service.
(Author.)

are very expensive when volumes are limited, become more cost-effective than buses when ridership exceeds 50,000 trips per hour. This means that large, expensive mass transit systems are almost always confined to large cities.

One way for mass transit systems to reduce their operating deficits is called peak load pricing. This means charging higher prices when demand is highest, that is, during rush hour. The concept is used in many arenas: movie theaters, for example, charge higher prices when demand is high, such as weekend nights. While peak load pricing is economically rational, it is often not politically feasible, as commuters get upset when told they must pay more. The problem is compounded when ridership stays constant but operating expenses increase, necessitating an increase in bus and metro fares.

8.4 Zoning

One of the most widespread and influential ways in which governments shape urban areas is zoning, which means legally permissible land uses in specific areas. Zoning specifies what type of development is allowed, and what is not, and therefore greatly constrains the decisions of real estate developers. It is by far the most common type of land use control. Zoning divides cities into zones of specific land uses (Fig. 8.9). Today, almost every city in the United States (except Houston) has zoning.

The United States does not have a national land use plan, so most such decisions are made at the local level, although federal environmental restrictions are important. State laws and state courts are also critical. Most zoning ordinances are enacted at the county or municipal level.

Zoning has an important purpose: to minimize the damage caused by market failures, or what economists call negative externalities, or negative welfare impacts. An externality occurs when one person's decisions affect the welfare of another, for example, if a developer's new building blocks a person's view and causes the value of that parcel to drop accordingly.

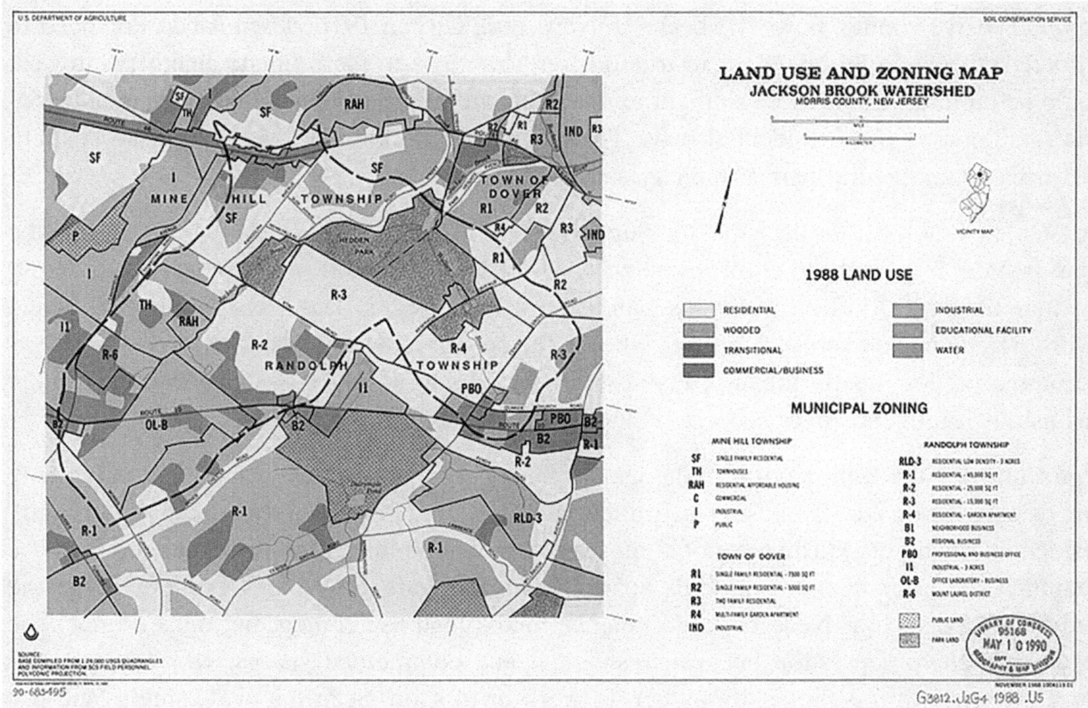

Figure 8.9
An example of a municipal zoning map. Map lines delineate study areas and do not necessarily depict accepted national boundaries.
(https://commons.wikimedia.org/wiki/File:Land_use_and_zoning_map,_Jackson_Brook_watershed,_Morris_County,_New_Jersey_LOC_90683495.jpg.)

Zoning exists to separate incompatible land uses, in which one type of use negatively affects neighboring parcels. For example, a slaughterhouse and an elementary school are incompatible, as are a hazardous waste dump and a playground. Noisy land uses such as fairgrounds are incompatible with residential areas.

Critics often allege that zoning is undemocratic because it allows unelected planners to control the decisions of private land owners. In this view, zoning commissions infringe on individuals' rights. At times, opposition to zoning can be heated for this reason. Notably, the most intense opposition often arises from those with the most to lose: developers. But supporters of zoning counter that zoning provides freedom from noxious land uses, and raises property values by reducing the risk of noise, congestion, or pollution. In this view, zoning is essential to protecting the quality of urbanites' lives.

Zoning has a long and checkered history. Premodern cities used it from time to time. In the 19th century, it was used as a means to enforce racial segregation. In 1885, Modesto, California, used zoning to limit Chinese laundries to that city's Chinatown. Modern, comprehensive zoning, however, began in New York City in 1916, when it was deployed to enforce setbacks in buildings, or restrictions on how close to the property line office towers could be built, to enable more sunlight to reach the streets. It also began to separate industrial and retail spaces from residential ones. The result was the familiar "wedding cake" pattern that became an integral part of the city's skyline.

In 1926, in a momentous decision, the Supreme Court ruled in *Euclid v. Amber Realty* that the state had the legal right to enforce zoning codes. The decision allowed the city of Euclid to separate residential, office, warehouse, and commercial spaces. The decision allowed zoning to proliferate quickly across American cities in the next decade, a period when US society was becoming predominantly urban. Today essentially all land in all cities is zoned, and zoning has as much influence over property values and uses as does the market.

Unfortunately, the zoning norm in the United States is single land use zoning, in which only one designated type of land use is permitted in an area (e.g., low-rise residential, multifamily residential, retail, etc.). This type of zoning is less common in Europe, where mixed land use zoning is the norm (hence one finds apartments above cafes or stores). Some cities in the United States, such as New York, do practice mixed land use zoning, but most do not. The result is a sharp separation between residential and commercial spaces, which essentially forces workers to travel, usually by car, to work or to shop. Seen this way, single land use zoning promotes inefficiencies in terms of time and energy use. Single land use zoning is partially responsible for urban sprawl and excessive driving.

Zoning occurs in different forms because different zoning laws serve different purposes. Some of the major types of zoning include:

1. Environmental zoning: This is a set of laws designed to protect natural habitats, such as wetlands, greenlands, flood plains, and open spaces. It may be used to conserve agricultural land and protect it from rural-to-urban land conversion. Environmental zoning may also be used for habitat preservation or to protect endangered species or particular ecosystems, or for noise mitigation. Increasingly it is used for flood mitigation and to combat climate change.
2. Residential zoning: This includes restrictions on the types of housing that can be built in given areas, including low-density, single-family homes and apartment buildings. In some places it may include setbacks (stating how far from the street a house must be located) or minimum lot sizes.
3. Commercial zoning: This form is designed to regulate spaces used for light and heavy manufacturing, warehouses, retail trade, and office buildings. At times it has been important in municipal downtown redevelopment projects. Occasionally cities will permit transfers of development rights, in which a sending area (e.g., a rural district) can sell its rights to a receiving area (e.g., downtown), forcing developers to pay extra for the right to build buildings in cities.

4. Historic preservation zoning: As the name suggests, this type is designed to preserve historic architecture and districts, often those with esthetic importance. It is used to conserve very old buildings of historical significance, or neighborhoods in which there are numerous older houses (e.g., brownstones) (Fig. 8.10).
5. Exclusionary zoning is zoning designed to keep a particular group, usually an ethnic minority community, out of a town or part of it. This form was once widespread but today is technically illegal, which does not mean it has gone away, that is, it does not exist *de jure* but does exist *de facto*. The most common forms are minimum lot size zoning, which specifies that each house must have a minimum lot size (e.g., half an acre or an entire acre), which drives up prices and keeps people with lower incomes out, or prohibitions on multifamily structures, which are most commonly used by people with fewer resources.

Figure 8.10
These brownstone residences in Boston benefit from historic zoning.
(https://commons.wikimedia.org/wiki/File:Brownstone_row_houses_in_Boston,_Massachusetts_LCCN2011634241.tif.)

Exclusionary zoning shows that zoning can be used against as well as for people, and that zoning laws are always tied up in the politics of class and ethnicity.

For zoning to be effective, it must be based on ethics and principles that guide fair land use regulations. First, good zoning requires steady and continuous public input, such as hearings, to allow the voices of homeowners, renters, environmentalists, and developers to be heard. Zoning works best when it includes those whom it most affects, and public exposure helps to prevent arbitrary or discriminatory zoning. Second, zoning should be fair, that is, it should, in legal terms, prevent "undue hardship." Third, zoning is not static, but a changing set of practices over time. It should respond to the oscillations of the business cycle and state of the economy. For example, during economic downturns some ordinances may be relaxed to encourage development.

Frequently, exceptions to zoning ordinances are passed called variances. A variance is not the same as rezoning a parcel. In theory these are to prevent unfairness, although in reality the vast bulk (80%) are passed at the behest of developers. Poor zoning takes years to overcome, and may result in ugly, haphazard landscapes with considerable congestion.

8.5 The State and Housing Markets

Yet another way in which the state shapes cities is through housing. Governments play important roles in affecting the supply and demand of housing, and therefore its price, availability, and location.

In the 19th century there was essentially no state intervention in housing, leading to widespread housing shortages for the poor and squalid tenement buildings often populated by immigrants. Significant government involvement began during the Great Depression of the 1930s, when homelessness surged. For example, the 1937 Housing Act created the Federal Housing Administration, which offered subsidies to local housing authorities and mortgage insurance for homeowners. The Federal Housing Act of 1949 saw a sweeping increase in the federal government's role, providing funds for slum clearance (Chapter 3) and funding for 800,000 public housing units. Under the guise of "urban renewal," the federal government encouraged the destruction of enormous numbers of low-income housing units, inhabited mostly by working class minority groups, to pave the way for commercial developments. In New York, this led to the construction of Rockefeller Center in Manhattan and the Cross-Bronx Expressway.

In the 1960s, as part of President Johnson's War on Poverty, the federal government began to construct large quantities of public housing. Public housing is widely misunderstood. Essentially it is an admission that the housing market cannot provide sufficient quantities of a socially necessary good. Public housing in the United States comprises only 2% of the total housing stock, in contrast to Europe, where it is often 20% or more. Compared to Europe, American public housing consists of warehouses of lower-income populations: the vast bulk of residents are on public assistance. Most stay in public housing for less than 10 years; the median stay is four years.

Unfortunately, many of the "projects," as large public housing complexes came to be known, were poorly designed and maintained. Most became dilapidated, with malfunctioning heating and cooling systems and elevators. Many became covered with graffiti and riddled with gangs and drug users. Understandably, these are not pleasant places to live. Even so, given the dire shortages of low-income housing, there are long waiting lists to gain entry to such places. Some projects became so notorious that they were torn down: the prime example is the famous Pruitt-Igoe projects in St. Louis, which were deliberately destroyed in 1972 (Fig. 8.11). Since then, most public housing projects have been built on a small-scale basis, although they often encounter NIMBY opposition (Chapter 5).

In addition to the federal government's role, municipal governments are also active in shaping local housing markets. One of these is through property taxes, which for most cities is the bread-and-butter of their budgets and the major means of financing public services. Property taxes vary depending on the value of the property, assessment rates, and changes in the tax

Figure 8.11
Destruction of the Pruitt-Igoe Projects in St. Louis in 1972.
(https://commons.wikimedia.org/wiki/File:Pruitt-Igoe-collapses.jpg.)

rate. Commercial and residential properties are taxed at different rates. Rising property taxes have led to widespread frustration among homeowners in many cities, and most states offer some kind of relief, such as limitations on tax rates on long-term owners with fixed incomes.

Cities typically offer an array of housing programs, most of which are aimed at helping people living in poverty. These include housing rehabilitation subsidies, rent vouchers, low interest loans, sweat equity programs (where a loan can be repaid by working on a property), foreclosures on tax-delinquent properties, emergency and homeless shelters, and fire and safety codes and inspections. Some subsidies are offered to low-income senior citizens.

One of the most contentious forms of state intervention in housing is rent control. Rent control began during World War II, when the federal government briefly limited increases in rents due to the severe wartime shortage of housing. Since then it has been used only by municipal governments, particularly in cities where renters are a large portion of the population and a politically powerful voting bloc. Many states prohibit it altogether, and rent control tends to be found mostly in relatively liberal states on the east and west coasts. While rent control, as an absolute prohibition on rent increases, is quite rare, most cities use rent stabilization, which allows for modest increases. Even so, rents in rent-stabilized units are typically well below the market rate.

Arguments in favor of rent control maintain that the market cannot provide housing for people living below the poverty line. For the private sector to build low-income housing, rents must be high enough to generate a competitive rate of profit, but high rents put units out of reach of low-income people. Thus, this view holds, rent restrictions are necessary to ensure a supply of affordable housing. Arguments against rent control hold that restrictions interfere with the supply and demand for housing (Fig. 8.12). Opponents hold that because rent-restricted levels

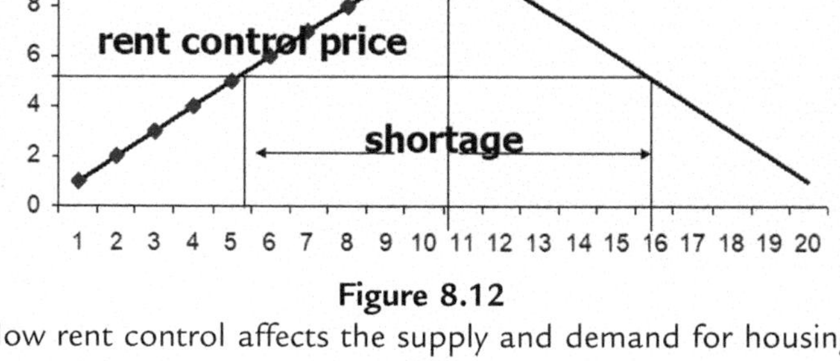

Figure 8.12
How rent control affects the supply and demand for housing.
(Author.)

are lower than the market prices, they lead to increased demand and insufficient supply, resulting in housing shortages. Rent restrictions, it is alleged, reduce incentives to developers to create new supplies, and punish newcomers to a city, as long-time residents of rent-stabilized units do not move out and free their units up to new tenants. Defenders of rent control note that housing shortages are common in cities without restrictions, and that without rent controls a wave of homelessness would materialize.

8.6 Chapter Summary

Cities are not simply the products of mythologized "free markets," but also of the state, or all levels and types of government policy. Governments' policies regarding cities vary widely around the world: in the United States, the federal government's role has been relatively small, whereas in many European social democracies the national government oversees land use controls. In all countries, the national government affects cities through the construction of a transportation infrastructure, airports, ports, water systems, housing policies, and poverty relief programs.

Urban planning, which had roots in the 19th century, came into its own in the early 20th century during the Progressive Era. Urban planners work behind the scenes, and their actions are often not visible to the public. Urban planning is critical in many respects: the design and layout of public services, such as schools and bus routes; economic development initiatives; building affordable housing; preservation of agricultural land; and protection of green spaces. Whereas traditional planning theory viewed planning as an apolitical realm consisting of detached experts, in reality urban planning is closely wrapped up with municipal politics.

Urban services shape the quality of life in urban areas. The most important is the public education system, which in many smaller cities is the largest budget item. Because schools in the United States are mostly financed through local property taxes, there are large variations in spending per student among school districts across the country. Urban services also include public transportation, such as buses and subways. Because the demand for these services is uneven over the course of the day, peaking during the morning and evening rush hours, for much of the time public transportation systems are underutilized.

Zoning is the legal limitation on land use. Like planning, zoning began in the early 20th century as a means of separating incompatible land uses. Today, every US city uses zoning to separate commercial and residential areas (single land use zoning), protect the environment, and preserve historic districts and agricultural areas. Exclusionary zoning, such as minimum lot size requirements, reveals that zoning can be used against some groups as well.

Finally, the state intervenes in urban housing markets. Because housing is a necessity, not a luxury, the state often confronts the issue of lack of affordable housing for low-income people. Public housing projects were one attempt to address this problem, but the large, multistory towers that were built in the 1960s were poorly designed and became dysfunctional warehouses for people with low incomes. Smaller public housing projects often confront NIMBY opposition. Rent control, adopted during World War II in many cities where renters are a potent political force, affects the supply and demand for housing: critics allege that rent control leads to housing shortages, whereas supporters note that there are housing shortages in cities without rent control as well.

Study Questions

1. Why is it wrong to claim that cities are only the products of markets?
2. How do national urban policies shape cities?
3. What are enterprise zones?
4. What is the rationale for urban planning?
5. How did Hausmann reshape Paris?
6. What are some things that urban planners do.?
7. How do urban public services differ from private ones?
8. Why does funding per student vary widely among school districts across the country?
9. Why is public transportation only cost effective at high volumes of passengers?
10. List four types of zoning and explain each.
11. Why are large public housing projects generally considered to be failures?
12. Does rent control contribute to housing shortages?

Further Reading

N. Bloom, F. Umbach, L. Vale (Eds.), Public Housing Myths: Perception, Reality, and Social Policy. Cornell University Press (2015).

C. Couch, *Urban Planning: An Introduction*. Bloomsbury (2016).

P. Hall, Cities of Tomorrow: An Intellectual History of Urban Planning and Design since 1880. John Wiley & Sons (2014).

R. Caro, The Power Broker: Robert Moses and the Fall of New York. Knopf (1974).

R. Caves, Planning in the USA: Policies, Issues, and Processes (5th ed.). Routledge (2023).

R. Weber, R. Crane (Eds.), The Oxford Handbook of Urban Planning. Oxford University Press (2015).

J. Jacobs, (reissue). *The Death and Life of Great American Cities*. Vintage (1992).

J. Levy, S. Hirt, Contemporary Urban Planning (12th edition). Routledge (2024).

M. Lewis, Who Is Government? The Untold Story of Public Service. Riverhead Books (2025).

J. Massengale, V. Dover, Street design: the secret to great cities and towns (2nd ed.). John Wiley & Sons (2025).

D. Merriam, The Complete Guide to Zoning: How to Navigate the Complex and Expensive Maze of Zoning, Planning, Environmental, and Land-Use. McGraw-Hill (2004).

CHAPTER 9

Urbanization in the Developing World

Chapter Outline

The vast bulk of the planet's cities, and urban inhabitants, are located in what is variously called the "Third World" (an obsolete term), the Global South (Fig. 9.1), or the developing world, including Africa, Latin America, and almost all of Asia (except Israel, Japan, South Korea, Taiwan, and Singapore). More than 80% of humanity lives in these regions, and they have higher population growth rates than do the developed countries of Europe, North America, Japan, and Australia and New Zealand. This is an enormously diverse set of societies, with varied cultures, religions, standards of living, and historical backgrounds. Most, but not all, were colonies of Europe at one point or another, which had tremendous implications for their social and economic development and role in the world economy. This diversity makes it difficult to generalize about cities in countries as varied as, say, Argentina, Mali, and Bangladesh. Nonetheless, the overriding feature of the developing world is relatively low incomes, as reflected in Fig. 9.2. Frequently, the rate of urbanization worldwide is associated with national incomes: wealthier countries tend to be more urbanized.

Today, most of the world's largest cities are located in the developing world (Table 9.1). These include Mexico City, Mumbai, São Paulo, Shanghai, Kolkata, Dhaka, Buenos Aires, and Karachi, among others. These megacities often have more than 20 million residents apiece, and are often growing rapidly. However, although they are huge in size, these urban areas are not global cities (Chapter 5) in that they are not typically large financial centers, do not have high concentrations of transnational corporations, and often play peripheral roles in the world economy.

This chapter explores the nature of urbanization in the developing world in several steps. It starts with a note as to why urbanization in the developing world is quite different from that in

Urban Geography. DOI: https://doi.org/10.1016/B978-0-443-30281-7.00009-8

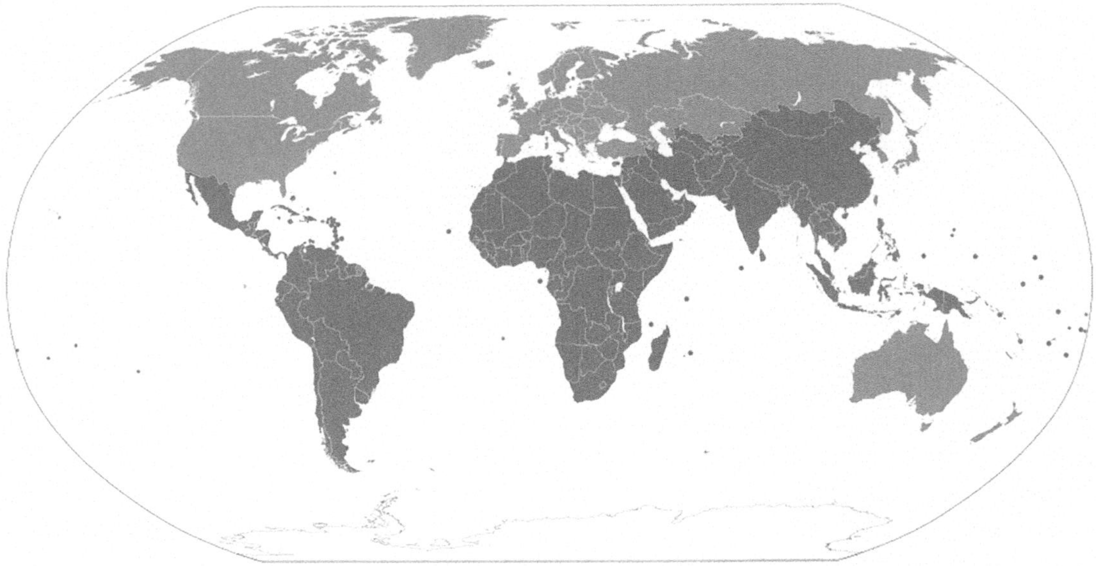

Figure 9.1
The Global South, or developing world, almost all of which was colonized by Europeans. Map lines delineate study areas and do not necessarily depict accepted national boundaries.
(https://commons.wikimedia.org/wiki/File:Global_North_and_Global_South.svg.)

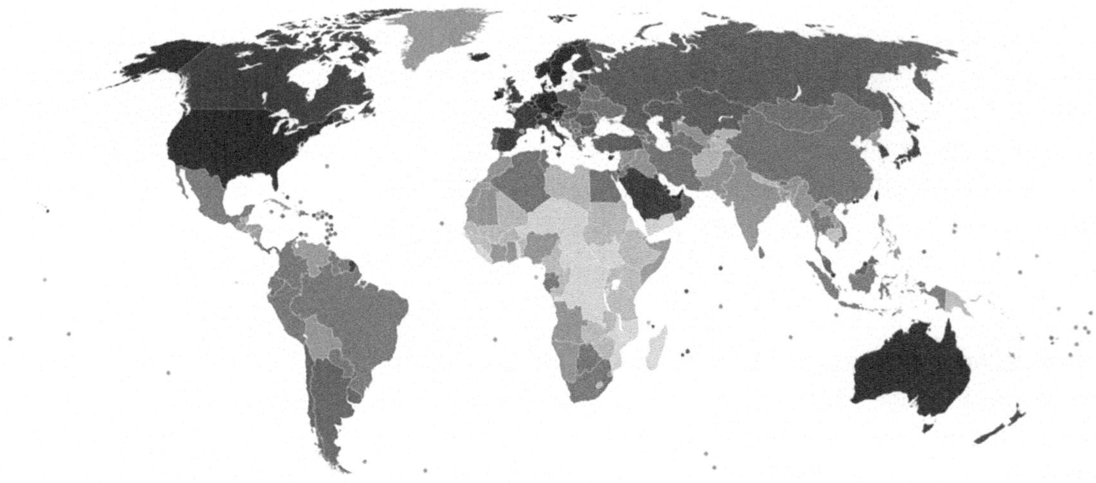

Figure 9.2
World map of GDP per capita, 2020. Map lines delineate study areas and do not necessarily depict accepted national boundaries.
(https://commons.wikimedia.org/wiki/File:Countries_by_GDP_(PPP)_per_capita_in_2020.png.)

Table 9.1 World's largest urban areas, 1970–2020 (millions).

Rank	Agglomeration, Country	Pop. 1970	Pop. 2000	Pop. 2010	Pop. 2020
1	Tokyo, Japan	26,615	34,450	35,676	36,371
2	New York, USA	15,880	17,846	19,040	19,974
3	Mexico City, Mexico	10,690	18,022	19,028	20,189
4	Mumbai, India	7,082	16,086	18,978	21,946
5	São Paulo, Brazil	9,614	17,099	18,845	20,544
6	Delhi, India	4,426	12,441	15,926	18,669
7	Shanghai, China	7,326	13,243	14,987	17,214
8	Kolkata, India	7,888	13,058	14,787	17,039
9	Dhaka, Bangladesh	2,221	10,285	13,485	17,015
10	Buenos Aires, Argentina	8,745	11,847	12,795	13,432
11	Los Angeles, USA	8,926	11,814	12,500	13,160
12	Karachi, Pakistan	3,989	10,019	12,130	14,855
13	Cairo, Egypt	6,450	10,534	11,893	13,465
14	Rio de Janeiro, Brazil	7,557	10,803	11,748	12,775
15	Osaka-Kobe, Japan	9,844	11,165	11,294	11,365
16	Beijing, China	6,034	9,782	11,106	12,842
17	Manila, Philippines	4,999	9,958	11,100	12,786
18	Moscow, Russia	7,623	10,016	10,452	10,524
19	Istanbul, Turkey	3,600	8,744	10,061	11,177
20	Paris, France	8,558	9,692	9,904	10,007
21	Seoul, South Korea	6,808	9,917	9,796	9,740
22	Lagos, Nigeria	1,890	7,233	9,466	12,403
23	Jakarta, Indonesia	4,813	8,390	9,125	10,792
24	Chicago, USA	7,160	8,333	8,990	9,516
25	Guangzhou, China	2,673	7,388	8,829	10,414

wealthy countries. Second, it offers a brief account of their historical context, including precolonial urbanization and the enormous impact of European empires. Third, it focuses on the drivers of urban growth in these countries, which center on rural-to-urban migration. The fourth part looks at labor markets in developing world cities, which determine social and geographic patterns of incomes and opportunities. The fifth section turns to residential spaces in these places, particularly the vast slum or shantytown districts that ring most large urban areas. Finally, it summarizes developing world urbanization on a regional basis to drive home the fact that it varies widely; despite the commonalities among cities, there is no "one-size-fits-all" model of urban growth.

9.1 Developing World Cities Reflect Unique Social Relations

It is common to view cities in the developing world, where poverty and unemployment are often widespread, as "backward" or "primitive" forms of cities in the developed world.

To view cities in poorer countries this way is to deny their unique historical experience, particularly colonialism, and the role that the countries in which they are situated play in the contemporary world economy.

The uneven spatial development caused by capitalism on a global scale, which started with colonialism in the 16th century, generated societies that are qualitatively different from those in the postindustrial West. Europe, unlike, say, Africa, Asia, or Latin America, was never colonized by a foreign power. The West was *un*developed, but never *under*developed, that is, its economies and infrastructure were never deformed by a foreign power. In forging the global economy, colonialism drove hundreds of millions, or billions, of rural people off their land. This observation underscores that developing world poverty is not "natural" or inevitable: it was produced, over many centuries, leading to societies with vast amounts of surplus labor, low incomes, high unemployment, and limited opportunities for social mobility.

In contrast, developing world societies paid a heavy price for centuries of Western domination. The rural masses of Europe or the United States never faced mass dispossession such as that encountered by people in former European colonies, a process that has driven billions of people off farms and out of small villages toward swollen megacities. The sheer volume of rural-to-urban migration overshadows anything found in Europe or the United States. Today, cities in the developing world play very different roles in the global economy. None are global cities in the sense of being international financial capitals. The developing world relies heavily on capital investments from foreign sources, notably transnational corporations, which often repatriate much of their profits. Developing countries rely far more on imports than developed ones ever did. Population growth rates are much higher than they ever were in Europe or North America. Population densities in the poor urban cities are much higher than in the West. Finally, the slums that surround developing world cities are not some transient or temporary phase but a permanent part of their urban landscapes.

These differences are critical if we are to avoid simplistic stereotypes that impoverished developing countries exist in some "stage" that is "behind" the West. Rather, we should see the societies and cities of the developing world in terms of the processes that created these places.

9.2 The Historical Context of Urbanization in the Developing World

Cities in Africa, Asia, and the Americas had a long history of urbanization prior to the era of European colonialism, conquest, and expansion (Chapter 2). As we have seen, Mesopotamia, Egypt, the Indus River Valley, China, Mesoamerica, the Andes, and parts of Africa all witnessed thriving civilizations, many of which dated back to the Neolithic Revolution. Many of these cities were much larger than those found in Europe before the Industrial Revolution, including Guangzhou in China, Istanbul, and Tenochtitlan, the Aztec capital. In Western

Africa, cities such as Timbuktu played major roles in trans-Saharan caravan trading systems. The Arab Empire saw cities such as Damascus and Baghdad become important economic and cultural centers. It is simplistic and erroneous therefore to think that cities in the developing world began with European takeover when in fact they had long histories that predated it. This point is essential to avoid the idea that urbanization in the developing world is reducible to a Western footprint.

Moreover, because cities in the developing world emerged under different historical circumstances than did those in the West, they cannot be regarded as "backward" or "primitive" versions of those in Europe and North America. Centuries of colonialism deeply reshaped them. Their economic role in global trade and investment is markedly different. Their population growth rates are much higher than any ever experienced in the developed world. Not surprisingly, their labor and housing systems are also quite different.

These caveats notwithstanding, it is also true that colonialism played a huge role in shaping and developing world cities, much as it also reshaped their rural areas, trade relations, and political dynamics. As (Fig. 9.3) reveals, Europe conquered essentially all of Africa, most of Asia, and the entire Western Hemisphere, which was the process by which capitalism went global and created a world economy.

Colonialism's impacts are hard to overstate. The conquest of colonies incorporated them into a worldwide division of labor. Typically, this process involved turning them into centers of a primary economy concerned with the extraction of raw materials from the earth, including

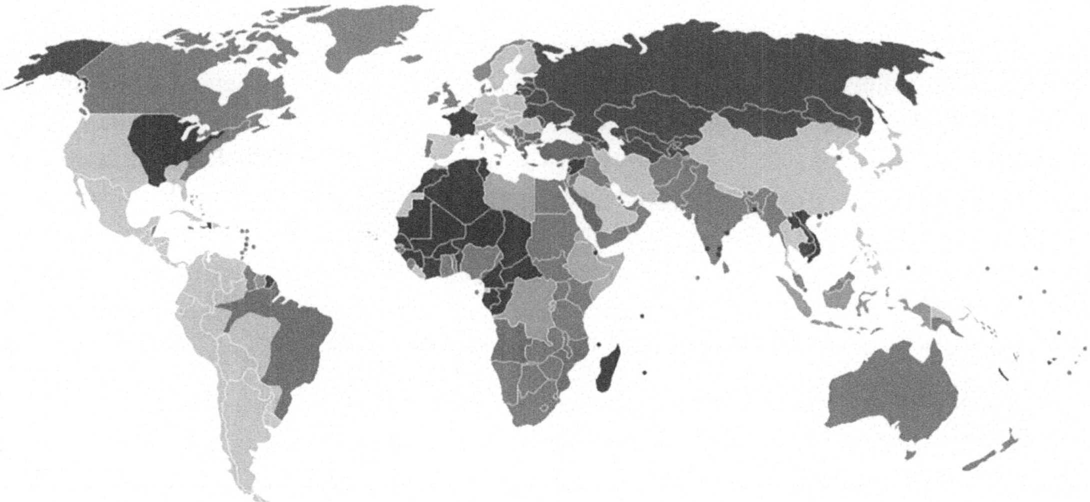

Figure 9.3
Map of European colonial empires at various historical periods. Map lines delineate study areas and do not necessarily depict accepted national boundaries.
(https://commons.wikimedia.org/wiki/File:EuropeanColonialism.png.)

logging, fishing, mining, and agriculture. The plantation system was the leading edge of the global commodification of agriculture. In the United States, this took the forms of tobacco and cotton plantations. In Latin America, Africa, and Southeast Asia, cash crops such as sugar, cotton, fruits, rubber, and tobacco were grown for sale abroad with the plantation system. Silver, tin, copper, gold, diamonds, and other ores were mined using the labor of enslaved people or peasants working in enslaved conditions. Mercantilist trade policies worked to suppress industrial growth in the colonies by reducing imports through stiff tariffs, such as the ones the British imposed on Indian textiles. This process is largely responsible for the fact that many developing countries today export low-valued goods and must import high-valued ones.

While some countries in the developing world had well-established traditions of urbanization before the Europeans came, such as the Arab world and China, in others, colonial powers played a major role. For example, the British constructed cities such as Calcutta (now Kolkata) in India, Rangoon (now Yangon) in Myanmar (formerly Burma), or Singapore (to control the strategically vital Strait of Malacca); similarly, the Dutch founded Batavia, Indonesia, which later became Jakarta. In China, colonialism took hold in cities such as Hong Kong and Shanghai, or "Treaty Ports" awarded after the Opium Wars of the 1840s. In South Asia, colonial port cities caused the population of coastal areas around Mumbai, Madras, and Calcutta (now Kolkata) to swell.

Ports were central to European maritime trade and control, and became important centers of commerce, often to the detriment of traditional capital cities further inland. In Peru, Lima displaced the Incan city of Cuzco; in western Africa, the famous trading city of Timbuktu declined as new maritime routes flourished; in India, coastal cities such as Calcutta, Bombay (now Mumbai), and Madras displaced the Mughal capital of Delhi; in Burma (now Myanmar), Mandalay gave way to coastal Rangoon (now Yangon); in Vietnam, the imperial capital Hue declined as the French colonial port city Saigon flourished; and in Indonesia, the traditional capital of Jogjakarta in central Java was marginalized by the Dutch port city of Batavia, later Jakarta. In Latin America, new port cities such as Havana, Buenos Aires, Santiago, Sao Paolo, and Rio de Janeiro also emerged. In Africa, Alexandria, Maputo, and Cape Town played a similar role. As cities flourished and offered more opportunities, millions of people left the poorer rural areas in waves of rural-to-urban migration. As a result, most of the world's people now live within 100 miles of the coast.

Often vestiges of colonial rule are seen in the architecture of cities, such as French colonial buildings in Hanoi, Vietnam (Fig. 9.4). Similar cases can be found in Mexico City and Bogotá, Colombia; indeed, many Latin American cities have plazas and large cathedrals that were constructed under colonialism. In Arab countries, French buildings can be found in Morocco and Algeria.

From the coasts, colonial railroads extended into the interior, often reaching into mineral-rich regions or plantations. In Africa, for example, rail lines extended to the interior from coastal

Figure 9.4
French colonial building in Hanoi, Vietnam.
(https://commons.wikimedia.org/wiki/File:French_COlonial_Architecture_-_Hanoi_-_Vietnam.JPG.)

outposts such as Dakar, Conakry, Abidjan, Lome, Maputo, Dar es Salaam, and Cape Town. The British built an enormous rail system in India, designed to expedite the exports of crops from plantations in the interior and to move soldiers around during military emergencies. A long-term consequence of this pattern is that the road and rail networks of developing countries often bear little resemblance to the distribution of the population that lives there; rather, they were constructed to facilitate the export of raw materials to the colonizing country, and today, to the global economy.

9.3 Drivers of Urban Growth in the Developing World

The developing world hosts large numbers of megacities (Fig. 9.5), often with populations greater than 10 million people. While there certainly exist large conurbations in the developed world, such as London, New York, Chicago, Los Angeles, Paris, Toronto, and Tokyo, these are overshadowed by the number of huge metropolitan areas in Latin America, Africa, and Asia. In addition, these large cities are growing much more quickly than those in the West ever did, largely due to high population growth rates and rural-to-urban migration. For example, Mexico City gains 22 people every *hour*, and similarly Sao Paolo gains 18, Kinshasa gains 63, Mumbai gains 51, and Delhi gains 79.

What is responsible for the large size and rapid growth of cities in the Global South? In part, their rate of expansion is a reflection of high birth rates: cities in the developing world are located in societies that often have higher rates of natural growth than those in the developed world, although birth rates are falling worldwide. But the major driver of the growth of

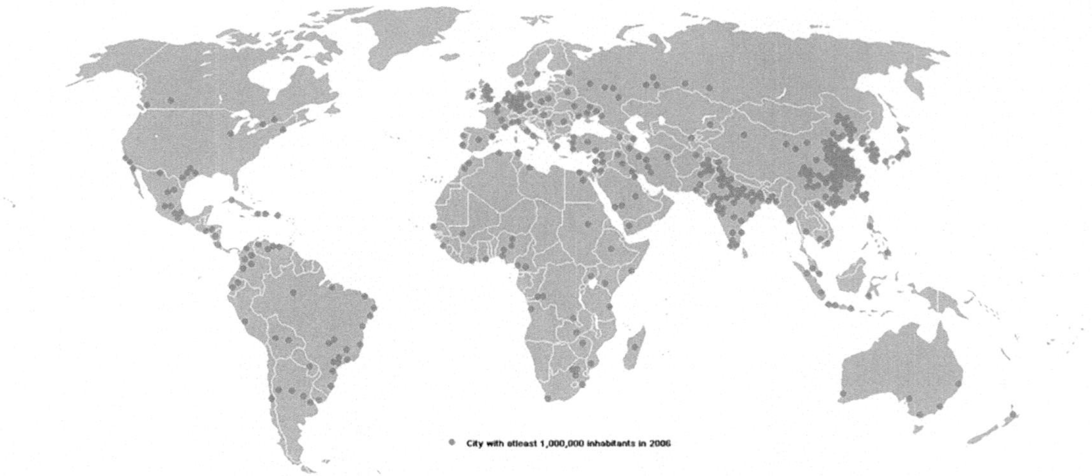

Figure 9.5
Map of distribution of world's cities with more than 1,000,000 people. Map lines delineate study areas and do not necessarily depict accepted national boundaries.

developing world cities is rural-to-urban migration. This was, of course, also important for cities in Europe and North America during the Industrial Revolution, notably in the 19th and early 20th centuries. Today, whereas some rural-to-urban migration still occurs in economically advanced countries, it is minuscule; in the developing world, in contrast, it occurs in enormous waves as many countries become increasingly more urban. In China, for example, between 1990 and 2020 more than 120 million people moved from the interior to the cities on the eastern coast, arguably the largest mass migration in history. Similar tidal waves occur in India, Mexico, Africa, Brazil, and Indonesia, providing a supply of labor that holds wages down to near-subsistence levels, creating a permanent poverty trap.

Rural-to-urban migration is a response to the different levels of opportunity, income, and standards of living between cities and the countryside. Rural areas are almost inevitably poorer than urban ones. Incomes and standards of living are lower than in cities. The differences in opportunities create "push" and "pull" factors that draw people into cities. The ongoing crisis of rural areas is one set of "push" factors. Rural areas typically have higher rates of unemployment than do cities, and young people there frequently encounter fewer opportunities to advance their careers. Young people tend to move more often than their elders, as they have the highest total economic returns they can expect to receive from migrating over future decades. In long distance moves, young men predominate. Highly unequal patterns of rural land ownership—a legacy of colonialism—mean there is relatively little arable land for farming families. High rates of population growth lead to declining amounts of arable land per person, putting pressure on young people to leave. Agricultural mechanization may enhance this problem by reducing farm jobs. Rural infrastructures are often poorly maintained, such as potholes in roads, driving up transportation costs. Access to health care in rural areas is often lacking, including clean drinking water. For those seeking higher education, moving to the city is a must. Finally, many rural people encounter ecological challenges, including soil erosion, deforestation, and landslides, the effects of which are being amplified by climate change. In short, the growth of cities in the developing world is inseparable from rural changes.

In contrast, cities offer a wealth of opportunities, both real and perceived. Urban job opportunities are relatively plentiful, unemployment rates are lower than in the countryside, and incomes and standards of living are higher. Access to education, health care, and transportation is relatively easy. Cities also offer better access to mass transit, the internet, electricity, entertainment, and other conveniences. Foreign-owned companies are far more likely to invest in cities, creating jobs there. Most public sector jobs are concentrated in cities. Many governments subsidize food in urban areas but not rural ones, in part as a way of ensuring cheap labor for transnational firms.

However, rural migrants operate under conditions of considerable uncertainty. Many are poorly educated and may be naïve, gullible, or overconfident. Low levels of reliable information mean that the decision to move is fraught with dangers. Some buy into mythologized accounts of urban life, leading to unrealistic expectations. Many such migrants are bitterly disappointed when they find that urban life can be as harsh and unforgiving as that in rural areas.

Rural-to-urban migration can be a complex process. Often it occurs through step migration, in which migrants slowly work their way up the urban hierarchy, moving from a rural hamlet to a small town, then to a small city, then to a larger city, and ending up in a huge megacity, usually in the slums and shantytowns. Also common is chain migration, in which migrants from a specific town or village may move and recreate their old neighborhoods in a large metropolitan area. Chain migration often occurs when a few migrant pioneers pave the way, sending information back to their friends and relatives in their home town. New migrants rely on earlier ones for information about jobs and housing. The result of chain migration is that parts of a city may speak a dialect from another part of the country.

Some migrants to urban areas may continue their journey and move overseas, finding work in North America or Europe. Some may obtain legal permission to do so, while others may be undocumented. The decision to move abroad is a difficult one, for it means leaving friends and family and undertaking what is often a perilous journey. Some are caught and deported back to their home countries.

Migrants who move to cities abroad frequently send part of their incomes back to their families in what is known as remittances. Examples include remittances sent by Latin American immigrants to the United States and African and Middle Eastern immigrants to

Europe. These are often an important source of income for families in rural areas. Those with relatives abroad may have more disposable income than those without. Remittances form a significant part of the Gross Domestic Product of many poorer countries.

9.4 Labor Markets in Developing World Cities

The world of work in developing world cities differs significantly from that in the West. Given the annual tidal waves of rural-to-urban migrants, the supply of unskilled labor is enormous. Incomes are, on average, much lower, as are standards of living. Unemployment rates tend to be persistently high. The structure of job opportunities is also quite different, as relatively few workers enjoy stable and well-paying jobs.

In the developed world, urbanization occurred on the heels of the Industrial Revolution, that is, it was synonymous with industrialization (Chapter 2). The firms and factories that emerged in cities generated a steady and substantial demand for labor. While developed world cities experience poverty, often severe, it rarely matches the severity of that found in the developing world.

In contrast, much of Latin America, Africa, and parts of Asia experience urbanization without industrialization. While large parts of the developing world are urbanized, typically underdeveloped countries lag behind the economically advanced countries in this regard. Today, about 56% of the world's people live in cities: for the first time in human history, the majority of the planet resides in urban areas. Not surprisingly, the proportion of each country's people who live in cities varies widely around the world (Fig. 9.6). In parts of Latin America, urbanization rates resemble those of North America or Australia, where 75% or more of the population lives in cities. In Asia and Africa, however, the proportions are generally lower. Only half of China lives in cities, and in wide swaths of Africa less than 20% do so.

While natural growth rates in cities in the developing world are higher than in the developed world, the growth of cities in less developed countries is primarily due to the massive number of rural-to-urban migrants, many of whom are displaced by agricultural mechanization, unequal land distribution, high unemployment, lack of public investment and services, low farm prices, natural disasters (earthquakes, hurricanes), and wars in rural areas. This pattern was also true of cities in the developing West during the Industrial Revolution. However, because few developing countries have generated the industrial job growth that the West did, the labor markets and employment conditions in the developing world are fundamentally different. In short, urbanization in the developing world and the crisis of rural areas are two sides of one coin.

Most cities in less developed countries experience large flows of migrants from rural areas as a result of low agricultural prices, inadequate public investment, rural unemployment, and lack of land reform. Many migrants are attracted to cities by the expectation of jobs and

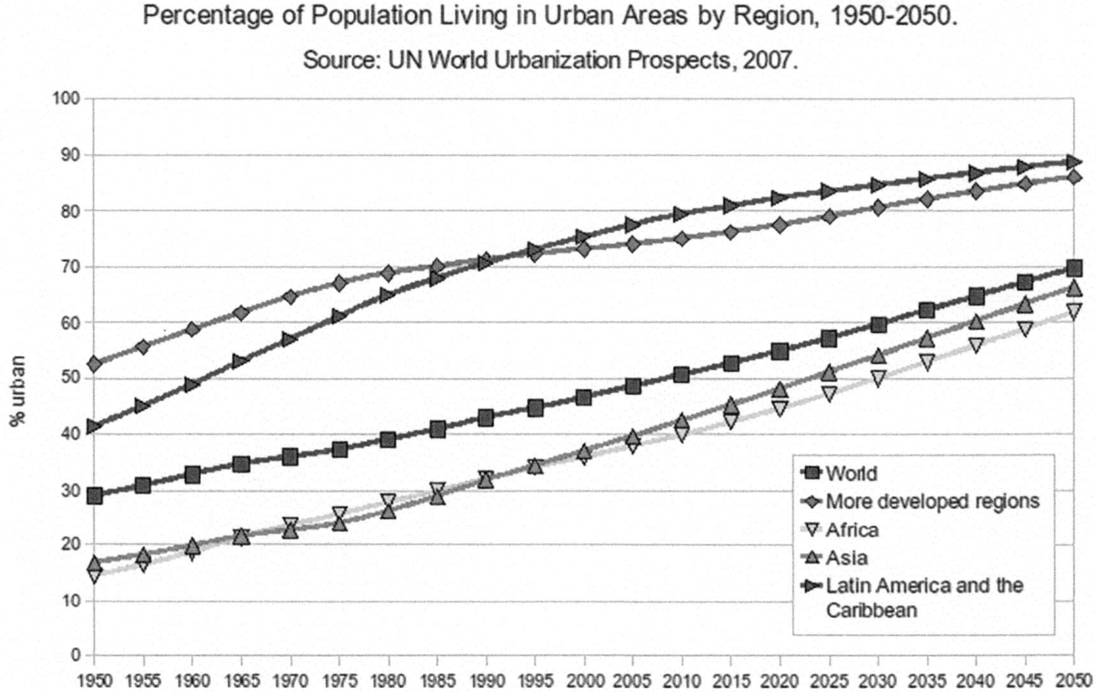

Figure 9.6
Urbanization rates over time in major world regions.
(https://commons.wikimedia.org/wiki/File:Percentage_of_Population_Living_in_urban_areas_1950–2050.png.)

higher incomes, expectations that may not be met in reality. Because many migrants move to the cities on the basis of their perceptions that there are greater opportunities to be found there (views that may be erroneous due to imperfect information), they often find themselves plunged into desperate circumstances. Once in the cities, many migrants cannot find work and contribute to the unemployment crisis. Other migrants find limited employment as handicraft peddlers or street vendors, often selling trinkets to tourists, forming part of the unregulated, low-wage, informal economy.

The more fortunate residents in developing world cities obtain jobs in the formal economic sector, in which employment pays relatively well and offers consistent working hours. Formal sector jobs tend to be registered with and regulated by the state. These are a minority of total employment. Formal sector jobs include those in government and registered businesses such as large corporations, including foreign-owned transnational ones. These include jobs in textile plants, such as Mexico's maquila factories (Fig. 9.7), electronics, automobile assembly, or the production of toys and many household goods. Formal sector jobs also are found in offshore oil drilling platforms. Formal sector, export-oriented jobs often tie developing

Figure 9.7
A maquila textile plant in Mexico, part of the formal economic sector.
(https://commons.wikimedia.org/wiki/File:Maquiladora.JPG.)

countries' cities to the global economy, such as the garment industry or electronic assembly plants—which, however exploitative they may appear to Western ideas, still offer higher wages than most local opportunities—tend to be the exception, not the norm.

Less developed countries' urban labor markets almost never generate sufficient employment opportunities for everyone, leading to high unemployment rates, or underemployment, in which migrants do not utilize their skills. Many find work in low-paying service jobs in the informal economy, which includes a wide array of jobs that are generally oppressive, low paying, and offer little financial security (Fig. 9.8). In many poor countries, the informal sector constitutes the bulk of employment opportunities, especially for women. (India, e.g., has 150 million women working in the informal sector.) Some informal sector jobs are small, family-run businesses, such as selling vegetables or cigarettes in a little stall or making clothes at home. Men may be employed pulling rickshaws, which is hard, exhausting work (Fig. 9.9). Child labor is common, such as carrying bricks at construction sites. Others find marginal

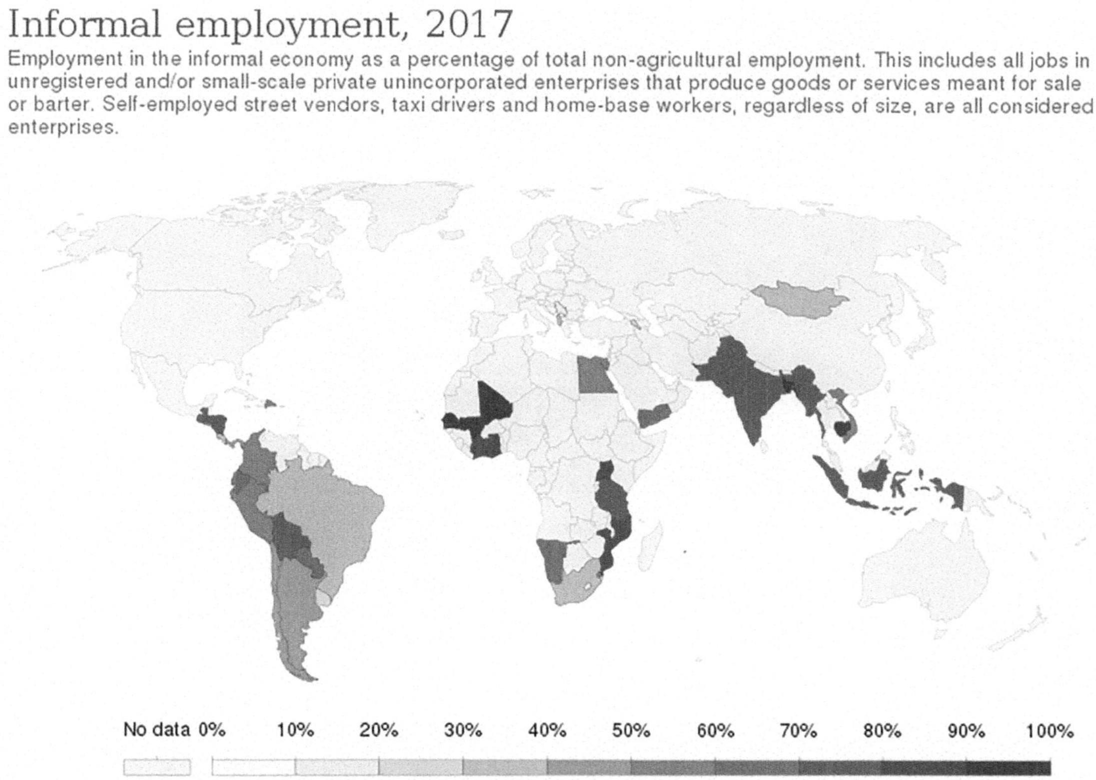

Figure 9.8

The extent of the informal economy across the globe. Map lines delineate study areas and do not necessarily depict accepted national boundaries.

(https://commons.wikimedia.org/wiki/File:Informal_employment,_OWID.svg.)

incomes selling trinkets and food on the streets (Fig. 9.10), prostitution, security guards, construction, delivery riders, selling illegal drugs, the black market, illicit currency exchange, poaching, or as casual day laborers doing construction. In Manila, the Philippines, for example, 30,000 people earn their living recycling garbage on the Payatas garbage dump (Fig. 9.11), as adults and children work together, stepping over debris to scavenge bits of plastic or tin cans. Children forced into labor in these conditions may not attend school and suffer from a lifetime of deprivation. Because such employment pays very little, informal sector workers tend to be vulnerable and impoverished. The informal sector flourishes when there are few formal sector opportunities, or when governments make starting a business difficult.

The existence and size of the informal economy have dire effects on the quality of life in cities. Whereas in the developed countries people with low incomes tend to be in the minority, often living in the city center, in the developing world, people living in poverty are usually a majority of the city's inhabitants, depending on the overall level of economic development.

Figure 9.9
Rickshaw drivers in India.
(https://commons.wikimedia.org/wiki/File:Rickshaw-Driver-Haridwar_AB.jpg.)

Figure 9.10
This street vendor in Mexico is typical of employment in the informal economy.
(https://commons.wikimedia.org/wiki/File:Vendedor_de_Dolores_Hidalgo,_Gto.jpg.)

Figure 9.11
The Payatas garbage dump, Manila, Philippines.
(https://commons.wikimedia.org/wiki/File:Payatas-Dumpsite_Manila_Philippines02.jpg.)

Low incomes mean that workers have less disposable income to spend, limiting job growth in sectors such as the retail trade and entertainment. Income taxes are low, starving local governments of funds. Most importantly, the resulting poverty means widespread desperation, with many people surviving hand-to-mouth. Despite widespread urban poverty, standards of living in cities still tend to be higher than in rural areas.

9.5 Residential Space in Developing World Cities

The residential geography of cities in developing countries differs from that in the developed world, a reflection of different labor and housing markets. Often, the relatively wealthy command the city centers, whereas people experiencing poverty live in the urban periphery. In many Latin American cities, for example, housing near the old colonial plaza in the center is relatively expensive, even gentrified (such as the Zocalo in Mexico City), while many live in slums on the urban periphery.

Billions of people today live in slums (Fig. 9.12). Slums, of course, were also common in Western cities during the Industrial Revolution (Chapter 3), but enormous areas of poverty have largely disappeared. In the developing world, where effective demand (demand backed by purchasing power) tends to be low, profits from the formal supply of housing are also low. Consequently, the construction industry that builds professionally designed and built houses is too small to satisfy the enormous demand for housing. As a result, many people must make

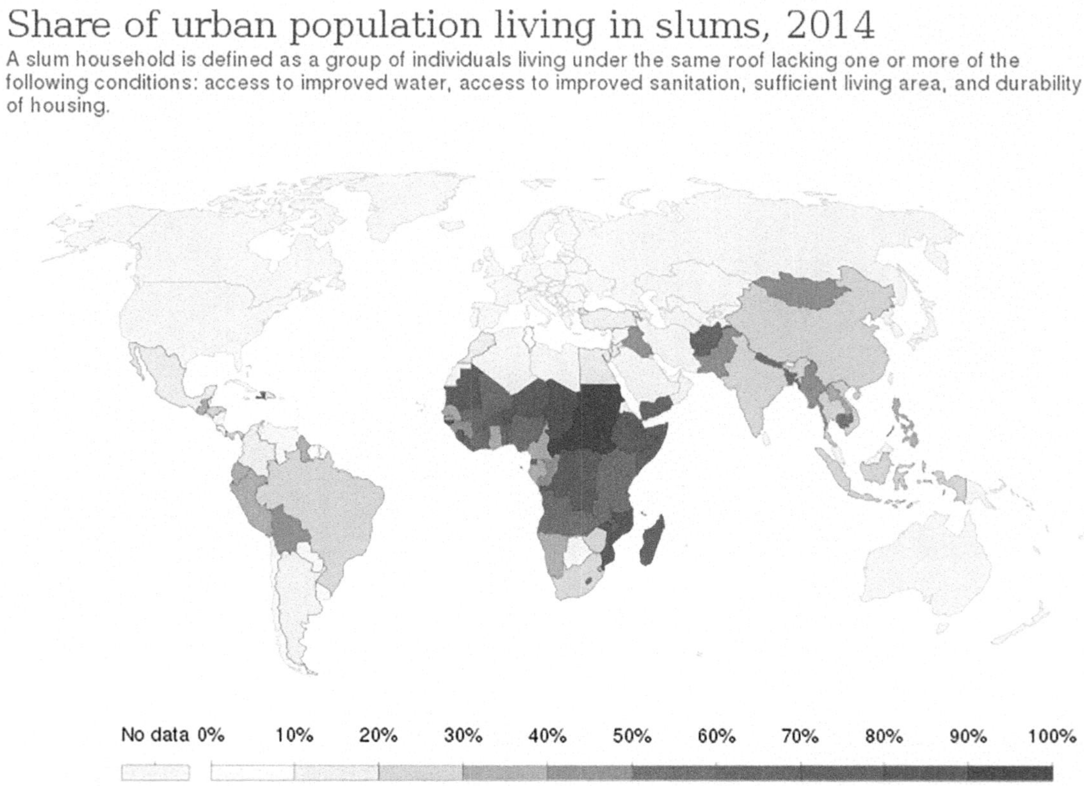

Figure 9.12
Proportion of urban population living in slums. Map lines delineate study areas and do not necessarily depict accepted national boundaries.
(https://commons.wikimedia.org/wiki/File:Share_of_urban_population_living_in_slums,_OWID.svg.)

their own houses, which are often small, dilapidated, and made from cheap materials such as cinder blocks, plastic, or sheets of corrugated tin. In some places, mud bricks may be used. Many urbanites inhabit squalid neighborhoods with substandard housing. Occasionally people may live in discarded trucking containers, in what is known as container urbanism.

These slums go by a variety of names: slums, Indian *chawls* or *bustees*, Indonesian *kampong* (Fig. 9.13), Turkish *gacekondu*, South African townships, Brazilian *favelas* (Fig. 9.14), and West African *bidonvilles* ("tin can cities") (Fig. 9.15). The Kibera slum outside Nairobi, Kenya, has more than 1 million inhabitants. In Mexico, slums are called *colonias*, some of which are also found in American states such as Texas and California. Most slums have a population between 30,000 and 100,000 people, but some can reach into the millions. Table 9.2 lists some of the largest slums in the world.

The structure of residential space is largely governed by the urban division of labor. Most people in the slums work in the informal economic sector: indeed, slums and the informal

Figure 9.13
Kampong ("neighborhood") in Indonesia.
(https://commons.wikimedia.org/w/index.php?search=kampong&title=Special%3AMediaSearch&type=image.)

Figure 9.14
Brazilian favela.
(https://commons.wikimedia.org/wiki/File:2019_Vidigal_Favela.jpg.)

Figure 9.15
Mathare slum in Nairobi, Kenya illustrates the housing of many poor people in the developing world. *(https://commons.wikimedia.org/wiki/File:Mathare_Slum_in_Nairobi_(5163712552).jpg.)*

Table 9.2 The world's most populous slums.

Slum Name	City	Country	Est. Population (2023)
Khayelitsha	Cape Town	South Africa	4,000,000
Kibera	Nairobi	Kenya	10,000,000
Dharavi	Mumbai	India	11,000,000
Neza-Chalco	Mexico City	Mexico	40,000,000
Orangi	Karachi	Pakistan	15,000,000
Tondo	Manila	Philippines	6,000,000
Cite-Soleil	Port-au-Prince	Haiti	3,000,000
Manshiyat Nasser	Cairo	Egypt	2,600,000

economy are both outcomes of one overarching process. Frequently, workers in the formal sector can afford to live in higher-quality housing than the slums, whereas people who have low incomes or are unemployed are confined to squatter settlements. In this way the divisions of the labor market are mirrored in residential space.

The living conditions in slums range considerably, from moderately safe neighborhoods to desperately poor ones. More than 1.5 billion people worldwide live in conditions that lack adequate housing, electricity, roads, transportation systems, schools, clean water, or sewers. (Notably, slum populations are often undercounted). Often densities in such places are far higher than in any developed country's cities. In many cities, more than three people live in each room. Privacy is rare. These homes are often uncomfortable, cold in the winters and broiling hot during the summers. The worst spaces do not have paved floors, which are often damp. Some do not have running water and rely on community pumps. Access to public toilets is limited, with dozens of families relying on one latrine, and they often malfunction. Garbage collection is typically inadequate, leading to the proliferation of rubbish that attracts rats. High unemployment, lack of access to health care, inadequate education and road systems all conspire to produce vast pools of suffering. Government assistance is either very limited or nonexistent. Fires may break out with devastating regularity and severity. Slums built on hillsides are vulnerable to landslides, which can kill hundreds of people in minutes. Government services such as health clinics, schools, and public transportation systems are typically inadequate. Many people walk to work or rely on public transportation such as buses. The police may act brutally and with impunity, robbing or murdering residents.

In countries with failed governments, in which law and order break down, gangs become the most powerful force in the slums: for example, Port-au-Prince, Haiti, is almost entirely ruled by gangs. Gangs may levy "taxes" through extortion and control expenditures in their neighborhoods. When the illicit drug trade flourishes, gang activity expands proportionately. The most vulnerable people are street children, including orphans, some of whom turn to pickpocketing or shoplifting to survive.

Many residents must pay rent to landowners, and frictions between the two groups are common. Without formal land titles, residents have little incentive to improve their living spaces. Commonly they lack the resources to do so. Slums may be built on vacant public spaces or places owned by wealthy landowners. When migrants seize buildings or land that belongs to wealthy landowners, the government may declare such squatter settlements to be illegal and bulldoze them.

The quality of life in slums, which varies, is often marred by overlapping health concerns. For people who are deeply impoverished, malnutrition is a chronic problem; in children this can result in stunting and delayed developmental growth. In places without sufficient running water and toilets, open defecation is common, particularly in India. Lack of clean drinking water and poor sanitation make such slums breeding grounds for infectious diseases. With poor hygienic conditions, slums are vulnerable to diseases like COVID-19, dysentery, malaria, cholera, and diarrheal diseases. Epidemics and pandemics sweep through slums with rapidity and regularity. AIDS is a constant threat. Skin diseases may be widespread. Air quality is often poor, particularly in homes with inadequate ventilation: worldwide, more than

1 million people die annually from polluted air caused by automobile exhaust, burning garbage, cooking with coal or wood, and the burning of fields after harvests, when the smoke blows into cities. Asthma, especially in children, is common. Extremely fine particulate matter penetrates deep into the lungs, causing respiratory problems. These factors contribute significantly to high mortality rates, particularly among infants, and low life expectancy.

Despite these frequently horrific living conditions, the reality is that the world's slums are slowly contracting in relative terms. Slums may grow in absolute population, but are declining as a share of the urban populace. For example, worldwide, slums dropped from 28% of urbanites in 2000 to 24% in 2020. This pattern reflects several structural forces. Severe poverty around the world is declining and incomes are rising, although unevenly over space. Malnutrition and famine have dropped significantly. As standards of living rise, more people can afford to live in better quality housing. In parts of the world that have experienced rapid economic growth, notably China and Southeast Asia, slums have become steadily smaller. In contrast, in places that have experienced little economic growth and high rates of population increase, slums remain firmly entrenched in the urban landscape. Finally, some governments and the World Bank have initiated poverty alleviation programs. For slums, this may include land titling programs to protect property rights. However, people living in poverty are not politically powerful, and often fall low in the list of government priorities.

9.6 Regional Variations in Urbanization in the Developing World

It is important to emphasize that given the enormous size and diversity of the developing world, urbanization there varies markedly. Cities in Latin America differ from those in Africa; cities such as Mumbai or Karachi are not the same as Jakarta or Tehran. Different urban forms reflect various historical experiences, the uneven impacts of colonialism, widely different cultures and standards of living, national urban policies, and the roles of countries within the worldwide division of labor. Nonetheless, there are also commonalities among these cities, the most important of which are their size, rapid growth, and widespread poverty.

East Asia has experienced the world's most rapid economic growth over the last 50 years of any major world region. Led by Japan, and now China, this region has seen incomes grow steadily, pulling billions of people out of poverty. Countries such as South Korea and Taiwan have escaped developing world status altogether. As a result, housing conditions have improved dramatically. With half of its population living in cities (700 million people), China hosts some of the largest cities in the world, including Beijing, Shanghai, Guangzhou, and Chongqing (Fig. 9.16). While China is not a rich country, its slums have declined significantly. The population of migrant workers lacking a *hukou*, or work permit, remains highly vulnerable and confined to the poorest parts of the cities. But overall, impressive gains

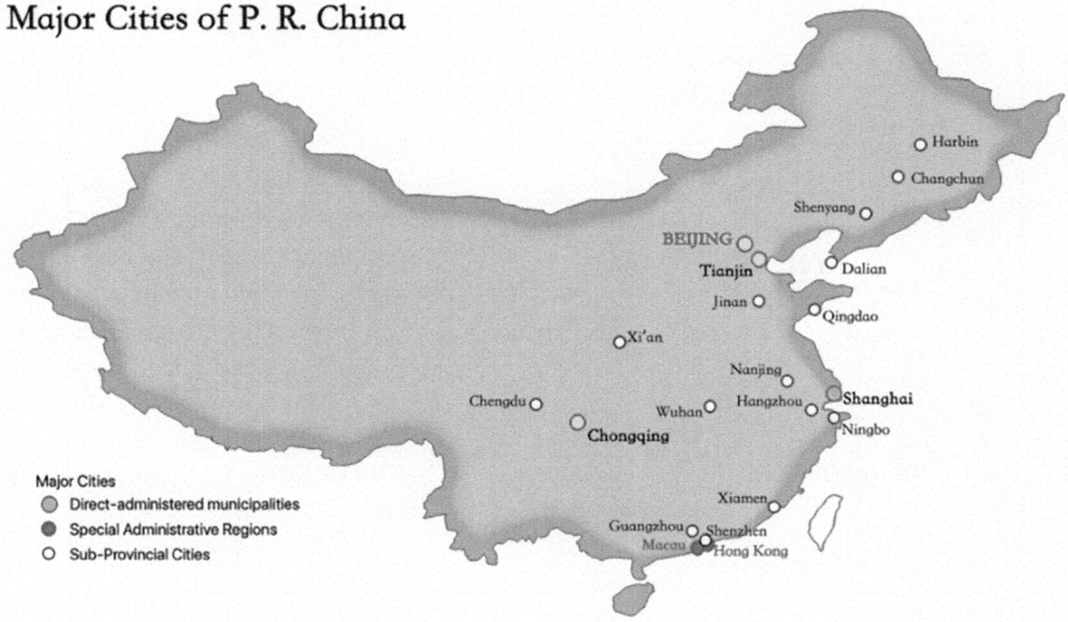

Figure 9.16
The largest cities of the People's Republic of China. Map lines delineate study areas and do not necessarily depict accepted national boundaries. *(https://commons.wikimedia.org/wiki/File:Major_Chinese_Cities_2020.png.)*

in China's economy have led more people to afford quality housing and even cars. Chinese cities suffer from the problems of rapid economic growth, including traffic congestion and air pollution.

In Southeast Asia, most of which is also growing economically, cities have exploded in size. Thailand has seen Bangkok become highly congested. Malaysia has become a middle-income country, as testified by the skyline of Kuala Lumpur, which is also the home of the Petronas Towers, some of the tallest buildings in the world. Singapore, which made itself into a prosperous center of finance and producer services, also exhibits startlingly beautiful architecture. In contrast, Indonesian cities like Jakarta and Surabaya have swollen to enormous sizes, surrounded by *kampong* slums. The poorer countries of Southeast Asia, like Indonesia and the Philippines, also exhibit gigantic slums.

South Asia remains the world's largest center of concentrated poverty. Even so, cities in India, Pakistan, and Bangladesh continue to grow rapidly. India's urban hierarchy is dominated by the "Big Four": Delhi, Mumbai, Kolkata, and Chennai (Fig. 9.17), each of which is surrounded by vast slums. Despite its widespread poverty, India also has a growing middle class, which fuels the growth of shopping malls. In Bangladesh, Dhaka is the

Figure 9.17

Major cities of India. Map lines delineate study areas and do not necessarily depict accepted national boundaries.
(https://commons.wikimedia.org/wiki/File:India_map_showing_major_cities,_parts_of_surrounding_countries,_and_Indian_Ocean.jpg.)

country's primate city, while in Pakistan Karachi (with 13 million people) is the country's largest urban center.

Latin America is the most urbanized part of the developing world. Most countries in the region are predominantly urban. Not coincidentally, the most impoverished nations are also the least urbanized, such as Haiti and Bolivia. Latin America also has some of the largest

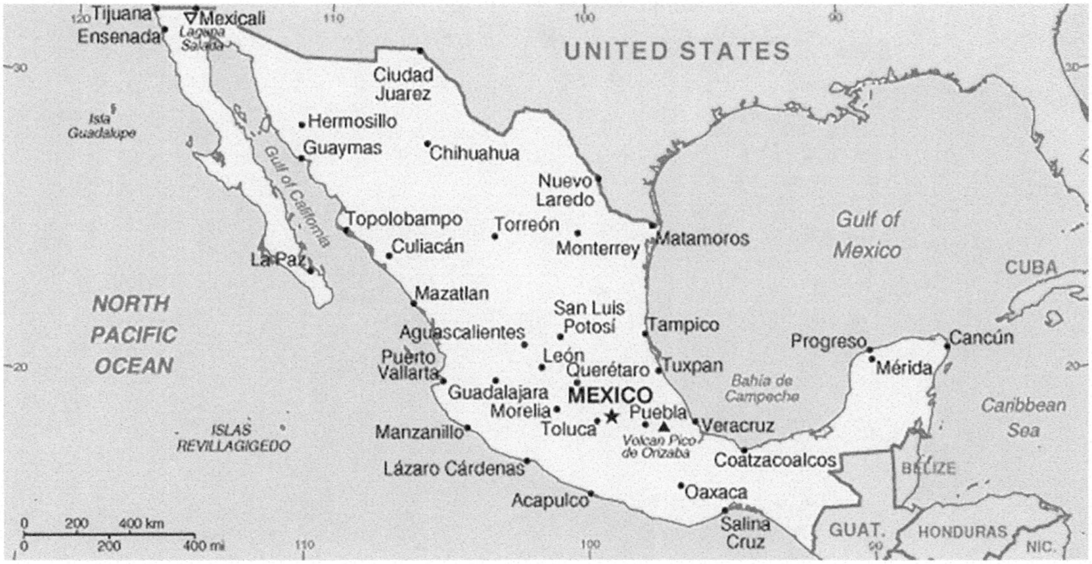

Figure 9.18

Major cities of Mexico. Map lines delineate study areas and do not necessarily depict accepted national boundaries.

(https://commons.wikimedia.org/wiki/File:Mx-map.png.)

conurbations on the planet, such as Mexico City (population 22 million), and Sao Paolo and Rio de Janeiro, Brazil. Brazilian cities exhibit *favelas* in which millions of people live. Many national urban hierarchies are dominated by large primate cities (Chapter 5) such as Guatemala City, Managua, Panama City, Havana, Lima, Buenos Aires, and Caracas. Many Latin American cities exhibit a plaza in the center, a legacy of Spanish colonialism, and enormous slums on the outskirts. In Mexico, for example, huge slum districts surround Mexico City, Guadalajara, Tijuana, Ciudad Juarez, Puebla, and Monterrey (Fig. 9.18). Income inequality is severe in most of Latin America, leading to enormous differences between luxurious upper class communities and the slums.

Sub-Saharan Africa is the world's poorest major region, and the least urbanized. Two-thirds of urbanized Africans live in slums. The slums of this region are growing at a rapid pace, 4.5% annually, or doubling every 15 years. The wealthiest country, South Africa, is also the most urbanized, with large cities such as Johannesburg, Durban, and Cape Town. The region is also home to gigantic urban areas such as Kinshasa, Lagos, Nairobi, Dar es Salaam, and Addis Ababa (Fig. 9.19). Due to its high rates of poverty, Africa has cities ringed by impoverished slum districts, such as the *bidonvilles* in Francophone West Africa.

Finally, cities in the Middle East reflect the enormous income inequalities of the region. Some urban areas, such as Cairo and Alexandria, are filled with large numbers of people living in

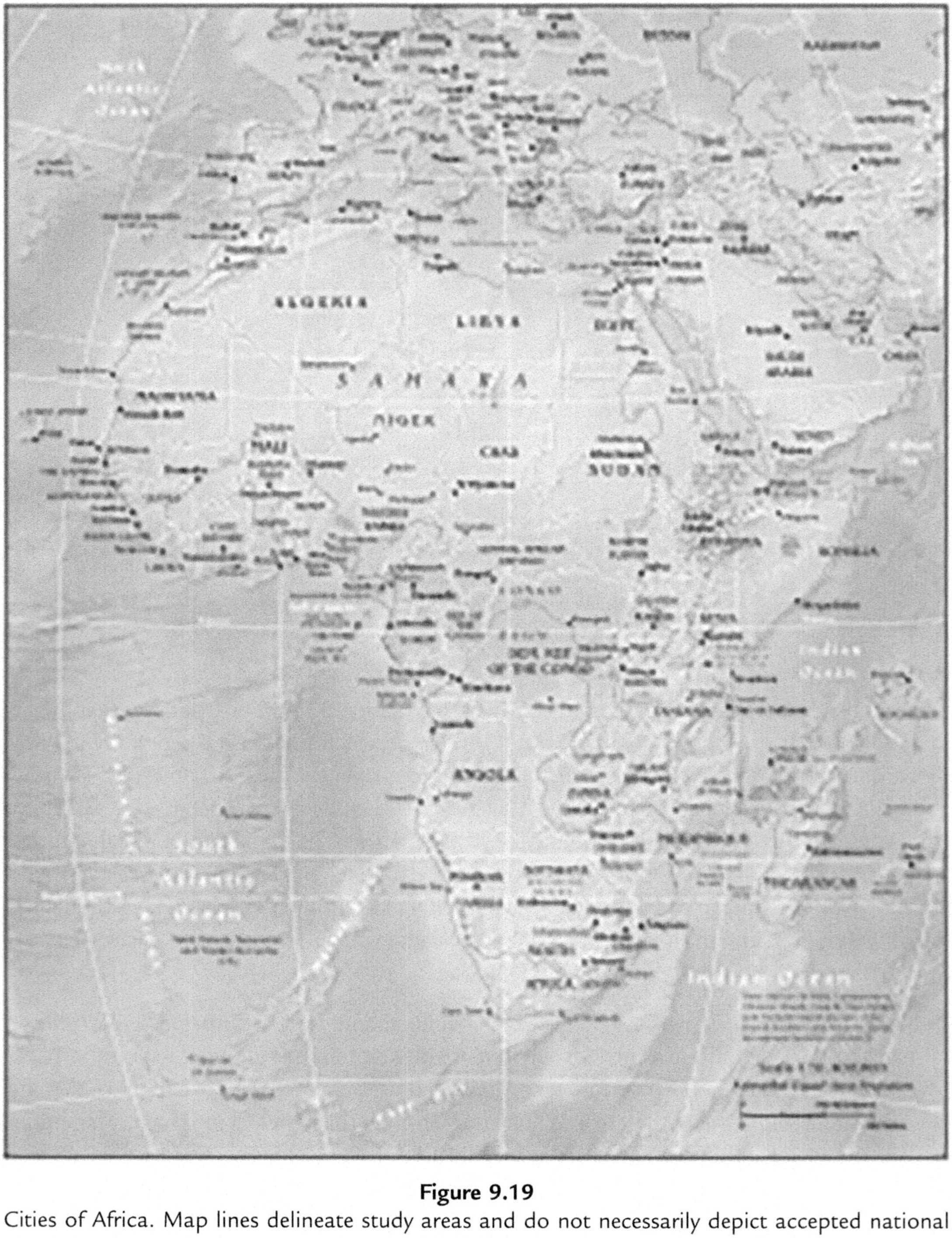

Figure 9.19
Cities of Africa. Map lines delineate study areas and do not necessarily depict accepted national boundaries.
(https://commons.wikimedia.org/wiki/File:Physical_Map_of_Africa.svg.)

Figure 9.20
The Burj Khalifa in Dubai, the tallest building in the world.
(https://commons.wikimedia.org/wiki/File:Burj_dubai_3.11.08.jpg.)

poverty. Israel, a middle income country, hosts relatively prosperous cities such as Tel Aviv and Jerusalem. Turkey is also a middle income country, and the *gecekondu* or slums on the outskirts of Ankara and Istanbul are slowly shrinking. Tehran dominates the urban hierarchy of Iran. The oil-rich sheikdoms of the Persian Gulf have created several spectacular cities filled with breath taking architecture, such as the Burj Khalifa, the world's tallest building, and the World Island, a series of luxury homes, in Dubai (Figs. 9.20 and 9.21).

9.7 Chapter Summary

Most cities, most people who live in cities, and the fastest urban growth rates are found in the developing world, the vast set of societies located in Latin America, Africa, the Middle East, and Asia. Roughly 80% of the human population lives in developing countries. There are enormous variations among, and within, countries in terms of standards of living as well as languages, cultures, and political systems. What almost all developing countries have in common are relatively low standards of living compared to the developed world and a history of European colonialism.

Figure 9.21
The World Island in Dubai.
(https://commons.wikimedia.org/wiki/File:Dubai_UAE.jpg.)

Developing world cities emerged under different historical conditions than did those in the developed world. There was, of course, a long urban tradition in places such as Asia, Africa, and parts of Latin America long before the Europeans arrived. But European colonialism had enormous impacts on these societies, restructuring them as part of a global division of labor designed to accommodate Western needs and priorities. Colonialism drew people to the coasts, and led to the dispossession of enormous numbers of peasants and farmers, resulting in waves of rural-to-urban migration and widespread poverty.

Cities in the developing world are growing much more rapidly than are those in the developed countries. In part, this growth reflects higher birth rates, although these have come down recently in many countries. The major reason is the migration of people from rural areas to cities in search of jobs, incomes, and better opportunities. Within developing countries, as well as at a global scale, migration is a response to uneven spatial development.

The labor markets that lie at the core of urban economies in developing countries are typified by a schism between the formal economic sector, in which jobs are reliable and pay relatively well, and the informal sector, in which employment is uncertain, unregulated, and pays poorly. The size of the informal economy is a measure of the well-being of a society, and it includes an enormous array of occupations ranging from street vendors to prostitutes, poachers to beggars, storekeepers to taxi drivers.

The structure of residential space in developing world cities reflects their relatively low incomes and the nature of the informal economy. Many, if not most, residents cannot afford housing constructed by the formal housing sector, in which professionals design and build houses. Rather, they must build their own homes out of whatever materials they can find. The result is the ubiquitous presence of slums and shantytowns on the outskirts of most developing world cities. These communities are often enormous, sometimes with millions of people, and while they vary considerably, most do not have adequate access to clean water, sanitation facilities, health care, and reliable electricity.

It is important to note that there are considerable variations in the nature, level, and type of urbanization in the developing world. Cities in a country such as India or Indonesia differ considerably from those in, say, Brazil or Uganda. There are marked variations in culture and political systems, as well as incomes. It is a mistake to treat the developing world as a homogeneous whole.

Many cities in the developing world contain buildings and spaces that resemble those widely found in the developed one, including luxury condominium towers, shopping malls, and event spaces like sports arenas and concert halls. The presence of these buildings, often in close proximity to destitute slums, reflects the sharp income inequalities that plague many developing countries: a relatively small but affluent middle class alongside large numbers of poor people. Moreover, because developing world cities are not uniformly poor, these types of spaces indicate that the simple dichotomy of "developed" and "developing" fails to do justice to the complexity of urban spaces and life. Indeed, given the steady rise of the global middle class and falling rates of poverty worldwide, some observers argue that the bifurcation between the developed and underdeveloped worlds should be discarded altogether in favor of a more nuanced categorization.

Study Questions

1. Why are cities in the developing world qualitatively different from those in the developed one?
2. Describe the impacts of colonialism on developing world cities.
3. How are the crisis of rural areas and the growth of developing world cities linked?
4. What push and pull factors drive rural-to-urban migration?
5. What is the informal economy? How does it differ from the formal one?
6. What are shantytowns or slums, and what produces them?
7. How are slums across the world changing?
8. Describe the regional differences in urbanization in Asia, Africa, Latin America, and the Middle East.

Further Reading

Neil Brenner, Debating planetary urbanization: For an engaged pluralism. Environment and Planning D: Society and Space. 0263–7758. 36 (3) (2018), 570–590, Available from: https://doi.org/10.1177/0263775818757510.

G. Bryan, E. Glaeser, N. Tsivanidis, Cities in the developing world. Annual Reviews Inc., United Kingdom Annual Review of Economics. 19411391. 12 (2020), 273–297. Available from: http://www.annualreviews.org/journal/economics, https://doi.org/10.1146/annurev-economics-080218–030303.

Tim Campbell, Alana Campbell, Emerging disease burdens and the poor in cities of the developing world. Journal of Urban Health. 84 (S1) (2007), 54–64, Available from: https://doi.org/10.1007/s11524–007–9181–7. 1099–3460.

M. Davis, Planet of Slums. (2007).

Erica Field, Property rights and investment in urban slums. Journal of the European Economic Association. 15424766. 3 (2) (2005), 279–290, Available from: https://doi.org/10.1162/1542476054472937.

J. Vernon Henderson, M.A. Turner, Urbanization in the developing world: Too early or too slow? Journal of Economic Perspectives. 0895–3309. 34 (3) (2020), 150–173, Available from: https://doi.org/10.1257/jep.34.3.150.

B. Marx, T. Stoker, T. Suri, The economics of slums in the developing world. Journal of Economic Perspectives. 08953309. 27 (4) (2013), 187–210, Available from: http://pubs.aeaweb.org/doi/pdfplus/10.1257/jep.27.4.187, https://doi.org/10.1257/jep.27.4.187United.

Alison E. Post, Cities and politics in the developing world. Annual Review of Political Science. 1094–2939. 21 (1) (2018), 115–133, Available from: https://doi.org/10.1146/annurev-polisci-042716–102405.

R.B. Potter, S. Lloyd-Evans, The City in the Developing World. Taylor and Francis, (2014), 1–244, Available from: https://www.routledge.com/products/9780582357419. https://doi.org/10.4324/9781315839370.

AbdouMaliq Simone, Cities of the Global South. Annual Review of Sociology. 0360–0572. 46 (1) (2020), 603–622, Available from: https://doi.org/10.1146/annurev-soc-121919–054602.

CHAPTER 10

The Social Geographies of Cities

Chapter Outline

Cities are not simply economic machines, nodes in the global economy, or destinations for migrants; they are also places of everyday life, where billions of people work, raise children, live, socialize, find entertainment, shop, and feel the full range of human emotions. To view cities without taking note of their social geographies is to reduce them to lifeless boxes.

Everyday life is a critical part of social analysis. In their daily lives, actors draw upon social rules and resources, which in turn structure their actions; hence, the structural qualities that generate social action are continually reproduced through these very same actions. The socialization of the individual and the reproduction of society and place are two sides of the same coin; the macrostructures of social relations are interlaced with the microstructures of everyday life. People reproduce the world, largely unintentionally, in their everyday lives, and in turn, the world reproduces them through socialization. In forming their biographies every day, people recreate and transform their social worlds primarily without meaning to do so; individuals are both produced by, and producers of, history and geography. Everyday thought and behavior hence do not simply mirror the world, they constitute it. History and geography are produced through the dynamics of everyday life, the routine interactions and transient encounters through which social formations are reproduced.

Everyday life is not a random set of interactions because people live in social worlds that are structured by broader economic, political, and cultural forces. This chapter explores the social

Urban Geography. DOI: https://doi.org/10.1016/B978-0-443-30281-7.00010-4

geographies of the city in several ways. It begins by examining class, which is often invisible yet hugely important. It then examines race and ethnicity and how they structure patterns of urban life. Third, it turns to gender and how it structures the urban experience. The fourth part explores sexuality as a force in shaping life in the city. Next, it focuses on age, as urban life varies considerably depending on how old one is. Sixth, it addresses the question of disability. The next sections explore religion, urban crime, the city as playground, and the city at night.

10.1 Class and Everyday Life

Class refers to the differential levels of economic and political power that different social groups have, that is, their incomes and access to resources. Class is often invisible; it is the least recognized category of social groups, but is very real nonetheless. People belong to a class even if they do not know it.

There are several definitions of class. According to Marx, the fundamental division in any society is between those who do and those who do not own the means of production. Under capitalism, this means that workers, who do not own the means of production, must sell their labor power for a wage to the capitalist class, which does own the means of production. Usually we define class by income levels, although the issue is more complicated than that. Class, however, is not simply an economic category, but also very much a political and ideological one; it shapes how people see the world, behave, consume, and live. Class is also a spatial phenomenon, reflected in different ways that varying levels of income and wealth play out across the urban landscape.

Class is visible in the urban landscape in many ways. High incomes and wealth afford the rich a greater array of residential possibilities, freedom from anxiety, and the ability to segregate themselves in enclaves such as gated communities (Chapter 6). Some wealthy people buy large homes in very expensive areas (e.g., Beverly Hills, California; Short Hills, New Jersey), including gated communities, where they are shielded from the world. These are populated by business executives, well-paid physicians, and celebrities. Manhattan, in New York, hosts "Billionaires' Row," where extremely wealthy individuals from around the world buy luxury penthouses (Fig. 10.1). A few cities have very high median incomes, such as Darien, Connecticut ($350,000) or Kentfield, California ($311,000). Cities like Aspen, Colorado, are synonymous with luxury, and expensive, which makes it difficult for low-wage workers there. Such places tend to boast upscale shopping outlets, expensive restaurants, great schools, art galleries, boutique hotels, and well-maintained public spaces. However, these neighborhoods are rare compared to the number of middle and lower class places.

Some relatively well-off professionals choose to live near the urban center, gentrifying older neighborhoods, to be close to jobs in producer services that agglomerate downtown and take advantage of its recreational opportunities. This process illustrates how economic restructuring leads to reconfigurations of urban residential space.

Figure 10.1
Billionaires' row, New York.
(https://commons.wikimedia.org/wiki/File:Billionaires%27_Row_2020.jpg.)

The middle class, under duress from stagnant incomes and rising prices, has fewer residential options. Their homes are filled with white-collar, salaried professionals with comfortable but not excessively high incomes, such as managers, lawyers, architects, accountants, teachers, sales representatives, software designers, and people employed in finance, real estate, and public relations. The great bastion of the middle class is suburbia (Chapter 6), in which 40% of Americans live. There are, of course, many different types of suburbs that range considerably in income. Some are quiet, leafy places, while others suffer from growing poverty. Suburbs tend to deploy single land use zoning (Chapter 8) and be dominated by single-family homes owned by their residents. Automobile usage is universal. Schools tend to be quite good. Such places tend to value privacy more than heterogeneity, communal spaces, or walkability. Suburbs tend to be predominantly White, although some suburbs have become home to a growing middle class of Asians, Blacks, and Latinos.

Towns, cities, and neighborhoods that developed around manufacturing acquired a distinctively blue-collar character, which now includes lower-paid workers in services. These tend to be populated by people who work with their hands and bodies, including plumbers, carpenters, mechanics, electricians, assembly line workers, construction workers, clerical workers, secretaries, customer service representatives, bus drivers, police officers, and lower level health care employees. Working class people employed in blue-collar jobs tend to live in smaller homes closer to the city center, as noticed by Ernest Burgess of the Chicago School a century ago (Chapter 6). With more modest incomes, home ownership rates tend to be lower. Schools tend to be passable but not great. These are people who shop at discount stores like Walmart. Some blue-collar cities have experienced massive gentrification and become expensive (e.g., Hoboken, New Jersey); some coastal fishing villages have experienced the same changes.

People with low incomes experience the city in very different terms. The vast majority of these people work, such as in low-paying jobs in the retail trade, restaurants, fast food, or as janitors, home health care aides, and in the gig economy. Some rely on public assistance. Constrained by low incomes, their residential choices are highly limited. Most do not own homes and must rent. Such neighborhoods are dense and crowded. Some can afford cars, others must rely on public transportation. With smaller houses or apartments, many people spend much of their time on the streets. Many live in states of financial precarity and have trouble paying their bills. Some experience episodic hunger. Those living in the most severe poverty are displaced (Chapter 6). Given how race and class are fused, low-income neighborhoods tend to be disproportionately populated by people of color. Schools tend to be poor, with high drop-out rates. And for those at the bottom of the social hierarchy, those without homes, daily life is a constant struggle to survive.

These examples illustrate how labor markets shape residential spaces in different ways. Varying income levels allow some groups a wide range of housing options while others are corralled into decrepit and dangerous communities.

10.2 Ethnicity and Everyday Life

Race and ethnicity—which are related, but not identical—are major dimensions in all societies today and hugely important in human geography. Race and ethnicity are social, cultural, political, and spatial constructions that demarcate and often naturalize differences among people, and are important foundations of identity. Race is typically defined as a biologically defined marker of difference among humans based on physical features such as skin color, facial features, or hair texture. Definitions of race typically privilege one feature, usually skin color, over others. Racism is the belief that distinct races exist and that some are superior to others.

Ethnicity is a term preferable to race because it highlights the social construction of culture and identity. Ethnicity reflects the cultures in which one lives, the traditions, customs, and lifestyles adopted by different groups. Ethnicity derives from both an internal sense of distinctiveness and an external perception of difference. The category exists to classify various groups of people based on specific social and cultural characteristics, with the most typical identifier being ancestry. Ethnicity is a social construction, where individuals are active agents in defining their ethnicity, at the same time that the category must be negotiated within a reactive, shifting environment.

The Chicago School of urban studies, which gave rise to urban geography as a discipline, initiated the first scholarly studies of ethnicity (Chapter 6), noting the formation of ethnic enclaves and theorizing how and why ethnic neighborhoods change over time. The Chicago School's theorization of ethnicity was primitive, however, and remained silent about the political dimensions of ethnicity, its ties to class and incomes, power, and social relations more broadly.

As we have seen in the examination of urban residential space, neighborhoods are often strictly segregated by race and ethnicity. Ethnic segregation is a significant fact of life in many large US cities (Fig. 10.2), in which African-Americans often populate large inner city ghettos surrounded by archipelagoes of White suburbs. The vast majority of people in American cities live in communities in which only one ethnicity prevails. The reasons for this include both the structural dynamics of labor and housing markets, which typically afford Whites advantages not available to most ethnic and racial minority groups, as well as cultural preferences, discrimination, and, at times, racism. The degree of ethnic segregation varies, of course, over time, and in some cities has gradually improved. Nonetheless, the existence of large, impoverished, minority-populated communities testifies to how class and ethnicity are closely intertwined in American society.

For African-American urbanites, daily life is often quite different from that of Whites. While Black people are a heterogeneous group, their incomes, on average, are lower, resulting in less purchasing power, illustrating how class and ethnicity are intertwined. Relatively few own their own homes. Many people are concerned about the quality of schools in their neighborhoods. Crime rates tend to be higher in Black communities than that in White areas. In poorer neighborhoods, many Black people live in fear of the police; there is a long, sordid

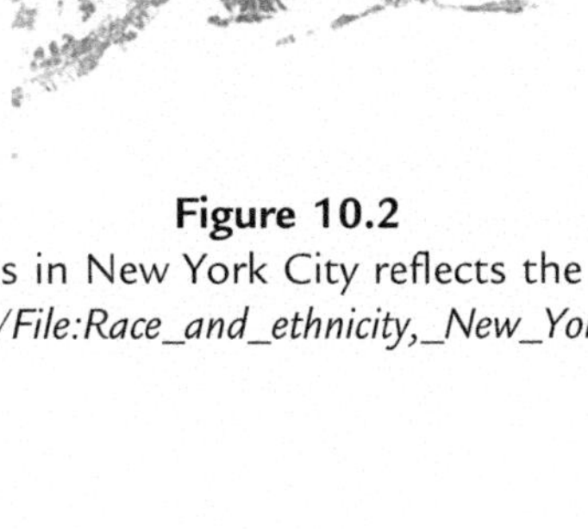

Figure 10.2
Distribution of ethnic groups in New York City reflects the high degree of segregation.
(https://commons.wikimedia.org/wiki/File:Race_and_ethnicity,_New_York_City_-_Flickr_-_Eric_Fischer.png.)

record of police abuse and murder of ethnic minority community members. When crime rates are higher, more people interact with the legal system, often in unpleasant ways. There are of course large, impoverished Black areas of many cities, which are the sites of public housing projects, gangs, and violence; but there are also middle-class Black neighborhoods, such as in Atlanta, Charlotte, Houston, Los Angeles, and New York, with well-educated populations of home owners.

Latinos or Hispanics are the largest minority ethnic group in the United States, comprising about 20% of the population. The Latino barrios of many cities, home to large numbers of Spanish-speaking residents, include both people born in the United States and immigrants, both legal and undocumented. Los Angeles, for example, has an enormous Latino community; East Los Angeles alone has more than 118,000 people, and English is rarely heard there. Latinos come from a wide variety of cultures, including Mexico, Puerto Rico, Cuba, Venezuela, and Central America, often speaking different dialects of Spanish. Unlike previous generations, the pressure to assimilate is minimal. Average incomes of Latino families tend to be low, and many are concentrated in blue-collar or low-paying service jobs. Unemployment rates tend to be high. The Catholic Church has a widespread following among Latinos and plays a significant role in shaping social life.

Asian Americans comprise about 5% of the US population and tend to be highly urbanized. Some families have historical roots in the Chinatowns that emerged during the 19th century (Fig. 10.3). Traditional Asian populations such as the Chinese and Japanese have been supplanted by more recent immigrants from Korea, the Philippines, and Vietnam.

The varied experiences of different ethnic groups reflect the enormous diversity of American society, which is reflected in urban landscapes. It is important not to stereotype groups or ignore their internal differences; people of the same ethnic heritage may have very different experiences depending on their class, gender, and age.

Given the residential segregation of ethnic groups, it is not uncommon for public spaces, notably public transportation systems such as buses and subways, to be the most integrated parts of the city.

10.3 Gender and the City

Gender is a fundamental dimension of human life, and revolves around the webs of masculinity and femininity that shape power, the allocation of resources, and the rhythms of everyday life. Masculinity is typically associated with power, strength, courage, and assertiveness, while femininity is often associated with beauty, passivity, empathy, and caring. Gender roles are social constructions, and are thus different from biologically given sex; while sex is relatively invariant across cultures, gender roles vary enormously historically and geographically. Gender is fundamental to everyday life, identity, and access to social

Figure 10.3
Chinatown in Manhattan, New York City.
(https://commons.wikimedia.org/wiki/File:Chinatown-manhattan-2004.jpg.)

resources, opportunities, and power. Gender does not simply consist of the differences between males and females in terms of outlook, identity, behavior, speech, movement, and opportunities, but also the forces that create male advantage and female disadvantage. Almost everywhere, today and throughout history, women live in a subordinate social position to men. More often than not, they work more, earn less, enjoy less social autonomy and fewer political rights, and culturally often garner less respect. Gender cuts across, intersects, shapes, and is in turn shaped by other lines of social organization such as class, age, ethnicity, sexuality, and geographic location.

Men and women often have different experiences of the city. Saddled with the bulk of the burden of childcare, and often cooking and cleaning, women tend to spend more time at home than men doing domestic work, which is unpaid, repetitious, and often boring. The jobs available to women outside the home are often low-paid and low-skilled, or "pink collar" jobs. Thus women dominate in minimum-wage jobs in retail trade, restaurants, clerical work, personal services, childcare, and nonphysician positions in health care. Women in the United States earn, on average, only 80% of what men earn, and the bulk of families below the poverty line are female-headed households, leading to the feminization of poverty. In the United States, large shares of very low-income households are headed by women. Black women are especially likely to be impoverished. Lower disposable incomes mean that the quality of life for women is often worse than that for men.

The growth of services of many types fostered rising rates of female labor force participation, including growing legions of women in professional occupations, although the common experience of a "glass ceiling" persists. As women entered the workplace, their changing roles reverberated to change gender relations at home, and raised issues for two-income families concerning childcare. Feminists challenged, for example, the common assumption in urban models of household commuting that the location of male jobs and the length of male commutes dominated the household location decision process. Moreover, the emphasis on male breadwinners and "stay-at-home moms" ignores the profound differences among women: Latina and African-American women, for example, have long worked outside of the home in greater proportions than White women. Others have focused extensively on gender differences in commuting, which can be substantial given that a disproportionate share of women work in part-time "pink collar" positions and assume the bulk of responsibility for childcare. Thus, feminist urban geographers have sought to ascertain whether women's generally shorter commuting times reflect their employment status or their domestic responsibilities.

The design of domestic space is strongly influenced by ideas about gender roles. The domestic sphere is thus coded as feminine, whereas public spaces are more often coded masculine. In many cases, including traditional Chinese and Islamic societies, areas for women and for men are strictly separated, and those for women are often isolated from the larger world. In Western societies, the most common domestic design has centered on the home as the domain of women and as a retreat from the larger world. Certainly, it has been usual in the Western world to locate the kitchen, designed as a separate area for the unpaid work of women, at the rear of the home. Typically women do the majority of domestic work inside the house while men do the bulk of outside chores. A home can be also be understood as a site of unpaid domestic work, oppression, and sometimes violence.

Gender affects other dimensions of urban life as well. Because women usually pick up children from school or childcare, they tend to have shorter commutes than do men. Nonetheless, burdened by work and domestic responsibilities, women have less free time than men do.

The morphology of the city may work to reinforce traditional gender roles and identities and serve as an obstacle to change. This is evident in shopping spaces such as retail stores and malls. Women are the principal clients at retail outlets, as witnessed by the proliferation of women's clothing stores. The gendered identity of consumers shapes how individual stores display merchandise, and more generally in advertisements directed at women.

For most women, the landscapes of fear, particularly fear of violent sexual assault, loom large; sadly, half the human race lives in terror of the other half. Women tend to experience fear most at night and in deserted places such as parking lots.

10.4 Sexuality and Everyday Life

Sexuality refers to a person's sexual orientation, and should not be confused with gender. Most people are heterosexual, but a significant number of people are not. Geographies of sexuality center on the idea that there is a spatial connection between various kinds of space, power, knowledge, and location.

Sexual behavior is not simply the product of some instinctual drive; rather, sexual activity is a multidimensional behavior that can be understood only in a larger cultural context. This is why a particular sexual behavior may be unacceptable in one cultural context and yet may be acceptable in another, and why cultural attitudes towards sexual orientation change through time. For example, in the early 20th century, the dominant view of homosexuality in the Western world was negative, a view reinforced by both the legal and medical establishments, but this has changed significantly a century later, although there is still much progress to be made.

Sexuality, or sexual desire and behavior, are typically very private matters, confined to the bedroom. But sexuality is also important in understanding urban life; it is a public as well as private matter. Debates over LGBTQ+ rights, for example, illustrate the politics of sexuality. The public nature of sexuality is also found in places frequented by young adults, such as bars, clubs, and entertainment districts, where the atmosphere is often highly sexualized, with public displays centered on bodies and clothing.

One of the ways in which sexuality is expressed in landscape, especially in large cities in the Western world, is through the gradual identification of a part of the city as being a residential and commercial area where the LGBTQ+ community dominates. Early examples include Greenwich Village in New York, the Castro district in San Francisco (Fig. 10.4), West Hollywood in Los Angeles, and Soho in London. Another expression of sexuality in landscape is a gay pride parade, march, or demonstration. These are now routine in many cities and are often major celebratory occasions attracting large tourist audiences. But they can be unsettling, and controversy often surrounds these displays of identity.

Sexuality is also evident in gay gentrification (Chapter 6). Usually childless, but not always, some LGBTQ+ people have more disposable income, and tend to be pioneers in urban areas beginning to experience gentrification. The LGBTQ+ community has taken the lead in the gentrification of many cities, including loft living.

Figure 10.4
San Francisco's Castro District, a famous LGBTQ+ neighborhood.
(https://commons.wikimedia.org/wiki/File:2017.04.27_SF_People_and_Places,_Castro_District,_San_Francisco,_CA,_USA_4088.jpg.)

10.5 Age and Everyday Life in the City

Age is an integral part of individual identity and behavior. The various stages of the family life cycle were recognized by the Chicago School, which emphasized its importance in, for example, Burgess's model of concentric rings. Burgess noted that young adults often live near the urban core to maximize their access to employment opportunities. When they have families, many people relocate farther from the urban center to procure more space for children. Finally, as they enter the postchild phase of life, many older adults return to urban apartments, often to be near health care facilities.

People's spatial needs and behaviors vary markedly by age, as do their experiences of urban space. For children, for example, the world consists of their home, school, and immediate neighborhood. Play is how mammals learn to socialize, and for children play is paramount (Fig. 10.5). The experience of childhood varies widely the world over among cultures, as well as by class, gender, and ethnicity, and has shifted over time. Whereas previously children spent most of their time outside on the streets or in parks, today they are primarily found indoors, drawn in by the popularity of movies and video games. As a result, childhood has become more regimented, less spontaneous, and less social in nature. For children from lower-income families, the city is often a dangerous landscape; gunshots are the leading cause

Figure 10.5
For children, the urban experience is largely one of play and exploration. *(https://commons.wikimedia.org/wiki/File:Anak_-_anak_bermain.jpg.)*

of death of children in the United States. Children from underrepresented communities often live in areas with poor air quality, resulting in high asthma rates.

As we grow older and we become more self-reliant, our spatial horizons expand to encompass wider areas. Riding a bike gives children much more spatial freedom. For teenagers, learning to drive is much the same, a rite of passage that offers them much more mobility.

For young adults, usually unburdened by the obligations of full-time work or raising a child, the city is largely a playground. Their daily lives tend to be structured around school, part-time jobs, and their social life. The creation and maintenance of social networks lies front and center. The quest for novelty, excitement, and adventure is important. Because sexuality and romance are prime directives at this age, this is the population that frequents bars and clubs. In such a context the digital and analog worlds become inseparable.

As young adults age, jobs and careers begin to become increasingly important. Many move within or among cities in search of employment. Living together to save on expenses is not uncommon for those just starting their careers.

For working-age adults, the city consists largely of daily routines. Saddled with the need to raise money to pay for housing, children, and other expenses, for many adults daily life consists of commuting to and from work and limited recreational time. Their class heavily affects their disposable income and freedom to structure time and space as they wish. So does ethnicity, as people of color tend to have less money than do White people. Adults in this group tend to be the most heavily involved in civic affairs, join NIMBY movements (Chapter 6), vote, or attend city council meetings.

Older adults experience the city in yet another way. In wealthy societies, with low birthrates and high life expectancies, the number of the middle-aged and older people is growing rapidly. Parts of the United States, such as Florida, have large numbers of older people and, thus, retirement communities. For those who have retired, dependence on public funds and services is critical. Some older people remain healthy, active, and vibrant until they pass away; indeed, an active social life is key to healthy aging. Many suffer from the infirmities that accompany old age, such as arthritis, which limits their mobility. Many older people choose to live as close as possible to health care facilities. Loss of the freedom to drive is a major blow to their independence. Some older people live alone and may be lonely; those with very low incomes rely on services such as Meals on Wheels. Some need home health care aides. The oldest and most infirm may live in assisted living facilities, which greatly narrows their geographical horizons; many are housebound.

10.6 Disability and the City

One out of six Americans lives with some kind of physical or mental disability. Physical disabilities take a wide variety of forms and levels of severity. Some people are born disabled; others become so through accidents, strokes, or disease. For disabled people with low mobility, the city is a challenging gauntlet of obstacles to be avoided, negotiated, and overcome (Fig. 10.6). Disabled people face landscapes that constantly threaten them with stigma, injury, and exclusion. Many are confined to home; for those in wheelchairs, ramps must be constructed. Trips must be planned in advance. Finding a restroom can be difficult. They often must undertake exhausting, circuitous routes that able-bodied people can shorten. Fear of crime prevents many from going to bus stops. Low-income and older disabled people are the most vulnerable. In the developing world, disabled people have a much harder time negotiating the city. Many disabled people face discrimination in employment as well. But people with disabilities are not defined by their disabilities; many live rich and fulfilling lives.

Some cities have taken steps to make the city friendlier to people with disabilities through inclusive environmental design. Buildings with stairs may be required to have elevators. Parking spots for disabled people are reserved for those with special permits. Buses may be equipped with ramps for lifting wheelchairs. Some lights at traffic stops beep to signal blind people that it is safe to cross. The Americans with Disabilities Act, enacted in 1990, sees the rights of disabled people as civil rights, and mandated that cities accommodate the needs of people with disabilities, with, for example, places for wheelchairs to cross curbs smoothly. Designing inclusive cities that facilitate daily life for people with disabilities reflects a worldview that values them as productive and important, embeds empathy into the urban landscape, and allows people with disabilities the same right to the city as everyone else.

Finally, the city is also home to people with mental illnesses of different types. Most people experience bouts of depression from time to time, which can be crippling. Other people have

Figure 10.6
Disabled people with mobility problems confront a different city than abled ones. *(https://commons.wikimedia.org/wiki/File:Kuo_Hua_Life_wheelchair_at_Keelung_City_Athletic_Meeting_20160521.jpg.)*

serious conditions like autism or schizophrenia, and are disproportionately represented among the unhoused. Some veterans suffer from posttraumatic stress disorder (PTSD), which contributes to being unhoused. Yet others suffer severe anxiety, such as from agoraphobia. Urban populations of people with mental illness began to grow rapidly in the 1980s and 1990s when the federal government cut back on social services and initiated the deinstitutionalization of psychiatric hospitals.

10.7 The Religious City

Religion also plays an important role in the social geography of the city. To what extent depends largely on how religious a particular group of urbanites may be and the nature of their faith.

Many cities became dotted with churches of different denominations of Christianity. Some acquired reputations for their prevalence, such as Brooklyn, New York ("city of churches") or Toledo, Ohio ("Holy Toledo!"). Christian churches today are often found alongside buildings serving other faiths, such as synagogues and mosques, and temples for Jews, Muslims, and Buddhists, respectively. However, the communities belonging to these faiths are small in the United States compared to Christianity, comprising only 2% of the population each.

For many urbanites, a church or similar building is an essential part of regular gatherings, a place to reaffirm their faith, and a location for socializing and participating in a community. Many churches perform charitable works, helping those in need, such as people who are hungry, those who are unhoused, and those who are unwell. Churches often run food or soup kitchens for people who are displaced. They are also important places of refuge during disasters or times of civil strife.

Compared to rural areas and small towns, which tend to be more traditional and religious in nature, cities are often seen as wellsprings of secularism. This is not to say that everyone in cities is an atheist; far from it. However, statistically, people who live in cities are less likely to believe in god(s) than are those in outlying areas. This notion contributed to the widespread view in the 17th through the 20th centuries that cities were amoral bastions of sin. Notably, even the United States, which is the only religious but economically developed country, has seen a slow decline in religiosity over time. This process has been reflected, for example, in the conversion of some churches into condominiums.

10.8 Crime and the City

For all the joys of city living, urban areas are also sites of criminality. It is important to see this issue factually. While large cities have a common reputation as being dangerous to live in, their property and bodily crime rates (not volumes) tend to be exaggerated. There is a significant mismatch between perceptions of crime and the reality of public safety.

The notion of cities as dangerous and crime-infested emerged during the Industrial Revolution, part of a broader constellation of cultural mores that viewed cities as sinful and sinister. The Chicago School (Chapter 6) was influential in studying urban crime as the result of alienation, social disorganization, and ethnic heterogeneity, but it tended to be simplistic in its interpretation; for example, Chicago School theorists held that urban density resulted in crime, which is simply not true. It is true that violent crime rates tend to be higher (roughly twice) in large cities as in rural areas. The most likely victims in both places are children. Moreover, crime rates rise and fall over time, reflecting many factors such as the state of the labor market, political tensions, and government policies. Governments tend to be particularly concerned when crime affects tourist areas. Some neighborhoods, especially in the suburbs, form Crime Watch groups to monitor outsiders and deter crimes.

Urban crime has many causes. The vast majority of crimes are committed by young men, whose behavior reflects, among other things, unemployment rates, wage levels, income inequalities, and restricted social opportunities. When formal opportunities and wages are low, more people will find crime to be attractive. Communities that are not cohesive and do not have strong social ties are more likely to experience crime. In such contexts, gangs will flourish. Some sociologists argue for the existence of subcultures of violence. But it is unfair to blame people experiencing economic hardship for crime when the vast majority never commit criminal acts; some types of crime, like white-collar offenses, are overwhelmingly committed by middle or upper-class White men.

The geography of urban crime rates varies considerably. They tend to be higher in low-income and ethnic minority communities. Criminologists identify "hot spots" where crime is much more frequent than normal. Some neighborhoods, for example, are much more prone to burglaries than others. Corners where drugs are sold are another case in point. Streets leading from entertainment districts to parking areas are likely sites of muggings. Dense neighborhoods with no off-street parking are most likely to see auto theft. Suburban homes occupied by dual-income couples are vacant during the day and thus attractive to burglars. People use their knowledge of crime hot spots in planning their daily routines, and the police often use the idea in their patrols.

Many cities have a history of episodic violent riots, often driven by grievances against the established governing order. There were frequent urban riots in the 19th century, driven by anger over many issues, including immigration, the military draft, food shortages, and labor disputes. Some riots are excuses for vigilantes to take the law into their own hands, including the grotesque lynchings of Black people that plagued many cities, especially in the South, during the late 19th and early 20th centuries. Some riots center on race, including attacks on ethnic minority populations, such as the Tulsa race riots of 1921 that destroyed a Black business community and led to the deaths of dozens of people. There were also riots against the Chinese and Latinos (e.g., the Zoot suit riots of Los Angeles in 1943). In the 1960s, many American cities were rocked by riots that followed in the wake of Black protests against racial injustices (Fig. 10.7). When the conditions of social reproduction deteriorate sufficiently—when many people have difficulty paying the rent or buying food—riots appear as a means of voicing frustrations. Riots are frequently met by the state with police or troops to quell disturbances. Riots remind us that the fabric of the city is tenuous and at times fragile.

Finally, cities are occasionally the targets of terrorist attacks. Terrorism is the use of violence against civilians for political purposes. Cities are attractive targets because terrorists can maximize the damage to lives and property. American cities have seen repeated terrorist attacks, including the 1993 World Trade Center bombing and the Boston Marathon bombing of 2013. Other countries have also experienced similar assaults, such as a terrorist attack on a hotel in Mumbai, India in 2008. In some cases lone terrorists have committed vehicular

Figure 10.7
Building burning during the Watts riots in Los Angeles, 1965.
(https://commons.wikimedia.org/wiki/File:Wattsriots-burningbuildings-loc.jpg.)

Figure 10.8
The World Trade Center after being hit by the first airplane impact, September 11, 2001.
(https://commons.wikimedia.org/wiki/File:Explosion_following_the_plane_impact_into_the_South_Tower_(WTC_2)_-_B6019~11.jpg.)

attacks against public gatherings and celebrations using trucks. Arguably the most famous urban terrorist attack was the September 11, 2001 incident in which Al Qaeda members hijacked jet airliners and crashed them into the World Trade Center in New York, killing roughly 3000 people (Fig. 10.8).

10.9 The City as Playground

Never before have so many people had so much time and money to spend on entertainment. Recreation and leisure are not simply diversions; for many people, they are the core of what makes cities worth living in.

Cities have long been synonymous with creativity, innovation, and fun. Their recreational districts include clusters of restaurants, art galleries, bars, museums, and coffee houses. Many postindustrial cities have culture-led renovations. The size and nature of urban recreational areas reflects their population size, incomes, transportation facilities, and local cultures. Many cities have "recreational hinterlands," or surrounding areas from which they draw people looking for fun. However, the privatization of leisure, such as televisions and the internet, threatened the viability of many public recreational spaces.

As disposable incomes rose over time, culture and entertainment became increasingly commodified. For those with spending power, cities can be enormous fun. Recreational spending is a major source of income for many urban retail outlets, restaurants, and theaters. Few places offer such dense concentrations of attractions as do cities. Healthy cities have a vibrant arts scene and varied culinary offerings. Large cities, especially those that draw numerous tourists, have museums. New York City is famous for its museums. Most cities have a wide array of restaurants and retail outlets. Some of these cater to wealthy people, such as Rodeo Drive in Los Angeles. Larger cities have famous concert and performing arts halls: Los Angeles has the Staples Center and Hollywood Bowl, New York has Madison Square Garden, Broadway theaters, and Radio City, and Washington, DC, has the Kennedy Center. Other recreational opportunities abound: zoos, botanical gardens, parks, historic sites and landmarks, monuments, casinos, riverfront walks (Fig. 10.9), conference centers, and amusement parks.

Figure 10.9
San Antonio's River Walk.
(https://commons.wikimedia.org/wiki/File:San_Antonio_Riverwalk_with_river.jpg.)

Recreational opportunities are often very important to the urban gentry, who gentrify neighborhoods, often at great cost to local working class populations. The notion of the creative class, or skilled white-collar professionals (Chapter 4), emphasizes that for time-stressed professionals, recreational opportunities are critical, particularly "authentic" ones that allow unplanned, serendipitous experiences. Pedestrian streets with boutique stores and restaurants loom large in this picture, part of an adult-oriented "experience economy" that does not fit into traditional family-friendly types of amusement. The surge in producer services, which has attracted large numbers of young professionals to urban cores and generated waves of gentrification, has resulted in the proliferation of many small recreational outlets. Often this has changed old warehouses and buildings into clubs and creative spaces. Many urban administrations appreciate these changes and alter urban planning to enhance and accommodate them.

Tourism, the world's largest industry in terms of employment, has also enhanced the importance of the city as a source of entertainment. Some cities are almost entirely given over to recreation, such as New Orleans and Las Vegas. Orlando, Florida, is dominated by Walt Disney World. Many US cities receive millions of visitors annually (Table 10.1). Occasionally cities may host spectacular events such as world fairs or the Olympics. A few cities (e.g., Venice, Amsterdam) experience "overtourism," in which the number of visitors exceeds the capacity of the local infrastructure to handle them, and have begun to place limits on the number of people who can come in.

Cities often invest heavily in recreational infrastructures, particularly sports arenas, as part of their place promotion (Chapter 4) efforts to lure visitors and tourists. Several cities have famous sports stadiums, such as Yankee Stadium (New York), Fenway Park (Boston) and Wrigley Field (Chicago), and the Superdome (New Orleans). Similarly, Arlington, Texas, hosts AT&T Stadium, New Orleans has the Ceasars Superdome, and Greensboro its

Table 10.1 US cities with the largest number of annual tourist visits, 2023.

City	Visitors (Millions)
New York	8.89
Miami	4.36
Los Angeles	3.60
Orlando	3.52
San Francisco	2.80
Las Vegas	2.08
Washington, DC	2.61
Chicago	1.41
Honolulu	1.32
Boston	1.15

Coliseum. Constructing new sports stadia is very expensive: examples include SoFi Stadium in Los Angeles ($5.5 billion), Allegiant Stadium in Las Vegas ($1.9 billion), MetLife Stadium in New Jersey ($1.7 billion), and Mercedes-Benz Stadium in Atlanta ($1.6 billion). Cities that attract large volumes of tourists must plan for sufficient hotel space and local transportation; in smaller cities, visitors can place enormous demands on water and sanitation systems.

10.10 The City at Night

Human beings are diurnal creatures, meaning they tend to rest at night. Consequently, cities during the nighttime are much quieter than during the day, and the streets are often deserted when most people are asleep. Children and older adults are rarely visible in the nighttime city. These patterns give a daily rhythm to life in the city, the ebb and flow of people in their daily existence. City rhythms are marked and accentuated by things like sleep-wake cycles, opening and closing hours of stores, and diurnal cycles of commuting. Generally speaking, the city becomes quiet at night; the streets are largely empty, and the sounds of traffic muted. For most people, the city at night is a quiet time of sleep and relaxation.

Of course, some people work at night, including staff in hospitals, workers in stores and all-night diners, as well as police officers, firefighters, janitors, taxi and Uber/Lyft drivers, and security guards. Some offices are open in the evening with people working late. Some people and occupations thrive during the night. Prostitutes conduct most of their business then. Gangs welcome the cover of darkness. Most violent crime in the city happens at night, especially in public spaces frequented by young men after consuming alcohol. These are members of the nighttime economy, which is much smaller than its daytime equivalent.

During the evenings, especially on weekends, public spaces of the city are dominated by young adults seeking leisure and socialization. This is when entertainment districts come to life, filled with restaurants, bars, pubs, clubs, and theaters. Greenwich Village, in New York, is one such example. Behavior in these contexts is often accompanied by conspicuous alcohol consumption. Nighttime playgrounds allow for

A time for transgression, a time for spending, a time for trying to be something the daytime may not let you be, a time for meeting people you shouldn't, for doing things your parents told you not to, that your children are too young to understand (Lovatt and O'Connor, 1995, p. 132).

However, after midnight such areas tend to empty out, despite the efforts of some entrepreneurs to hold late-night events (e.g., all-night raves, museum nights). Perhaps the most famous example is New York's New Year's Eve celebration held in Times Square, which draws enormous crowds.

Urban nocturnal playgrounds (Fig. 10.10) are not new: the introduction of electric lighting and gas street lights in the 19th century dramatically changed people's experiences of the night, making dark spaces more welcoming. Originating as a means of promoting law and order, street lighting opened up the spaces of the city at night into a public domain. Indeed, the city at night would be a practical impossibility without a well-functioning lighting system.

For many people, the city at night is a landscape of fear. Darkness has a long association with danger, with the inability to detect threats. The night is when most disruptive and antisocial behaviors such as public drunkenness, vomiting, vandalism, and public urination occur. Violent crimes tend to occur most frequently at night. Many, particularly women, do not venture from their homes out of fear; the urban night has traditionally been a dominantly male space. Most older people stay in at night. However, some people's fears of victimization are exaggerated and do not reflect the statistical risk of being exposed to crime. Some cities address this issue with better lighting, such as in parking lots, and closed-circuit televisions; the public spaces of many cities are under continuous surveillance. Gated communities at times adopt a fortress-like mentality that encodes fear materially, with gates, walls, guards, and cameras.

10.11 Chapter Summary

Cities are much more than nodes in the global economy or agglomerations of firms. Cities are where residents and visitors alike construct their daily lives, raise families, go shopping, and

Figure 10.10
Greenwich Village at night.
(https://commons.wikimedia.org/wiki/File:Greenwich_Village_at_night.jpg.)

have fun—in short, engage in social reproduction. But because societies are not simply random collections of individuals, but structured and patterned, everyday life in the city reflects the deep and profound imprints of dimensions such as class, ethnicity, and gender, all of which shape access to resources, power, worldviews, and identities. For those with sufficient funds, the city is a very different place than it is for those who are impoverished. White people experience cities differently than do ethnic minority communities who live in the shadow of discrimination, racism, or police abuse. Men tend to be outdoors more often than are women, have more money, fewer responsibilities, and more fun. Age makes a difference too: people use the spaces of the city in very different ways at different stages in the life cycle. Disability, too, matters, as negotiating urban space for people with mobility challenges is frequently inconvenient, exhausting, and frustrating. These experiences indicate that the urban experience varies widely, and there is no universal mode of living.

Cities have their dark side too. Urban crime rates are higher than those in rural areas, and pose a danger for economically disadvantaged people who are its most common victims. There is a long history of urban violence, much of it motivated by racism, that has taken the form of riots. People living in economic hardship, too, will erupt if social conditions deteriorate sufficiently.

Cities are also places where people enjoy themselves. Over the last half-century, as incomes have grown, the commodification of culture has led to a proliferation of recreational forms in the city, including shopping areas, entertainment districts, and tourist centers. For many urban areas entertainment and recreation are primary sources of income.

Finally, life in cities exhibits a variety of rhythms and tempos, most notable of which is the daily diurnal pattern of sleep-wake cycles. Cities are at their busiest during the day, when the vast majority of people are working, in school, or conducting their business. But beginning with the introduction of electric lighting, a gradual "colonization of the night" has occurred. For some, nights are the best time of the day. For many, however, the urban night is filled with fear.

It is worth emphasizing that there are considerable differences between the social geography and everyday life of cities in the developed and developing worlds. In wealthier countries, more people can afford relatively spacious homes, air conditioning, and automobiles. In individualistic cultures, people spend more time inside, watching television or playing video games. In many developing countries, in contrast, where homes are smaller, more people tend to spend their time outside, on the streets, socializing through face-to-face contact. Of course this blunt dichotomy is much more complex in the real world: there are people in New York who live like slum dwellers in New Delhi, and rich people in Kolkata who live like elites in Boston. How individuals structure their time, therefore, is the product of social forces ranging from the world-system and global economy to the dynamics of neighborhoods and individual locales.

Study Questions

1. How is class inscribed in the urban landscape?
2. Why do ethnic neighborhoods form?
3. How do men and women experience the city differently?
4. How is sexuality woven into urban geography?
5. How do residential needs change over the course of the life cycle?
6. What steps have cities taken to make life easier for disabled people?
7. What factors drive urban crime rates?
8. How has the privatization of leisure threatened urban public space?
9. What is the night-time economy?

Further Reading

L. Bondi, D. Rose, Constructing gender, constructing the urban: A review of Anglo-American feminist urban geography. Gender, Place and Culture. 10 (3) (2003), 229–245, Available from: https://doi.org/10.1080/0966369032000114000.

A. Collins, Cities of Pleasure: Sex and the Urban Socialscape. Routledge (2005).

M. Davis, Magical Urbanism: Latinos Reinvent the US City. Verso (2024).

Danielle van den Heuvel, Gender in the streets of the premodern city. Journal of Urban History. 0096–1442. 45 (4) (2019), 693–710, Available from: https://doi.org/10.1177/0096144218768493.

Therese Kenna, Neurodiversity in the city: Exploring the complex geographies of belonging and exclusion in urban space. The Geographical Journal. 0016–7398. 189 (2) (2023), 370–382, Available from: https://doi.org/10.1111/geoj.12512.

A. Lovatt, J. O'Connor, Cities and the night-time economy. Planning, Practice & Research. 13600583. 10 (2) (1995), 127–134, Available from: https://doi.org/10.1080/02697459550036676.

Setha M. Low, Urban fear: Building the fortress city. City & Society. 0893–0465. 9 (1) (1997), 53–71, Available from: https://doi.org/10.1525/ciso.1997.9.1.53.

Colin McFarlane, Fragments of the City: Making and Remaking Urban Worlds. University of California Press (2021). Available from: https://doi.org/10.1525/9780520382251.

Pia Christensen, Margaret O'Brien, Children in the City. Informa, UK (2003), 19–63, Available from: https://doi.org/10.4324/9780203167236–7.

Daphne Spain, Gender and urban space. Annual Review of Sociology. 0360–0572. 40 (1) (2014), 581–598, Available from: https://doi.org/10.1146/annurev-soc-071913–043446.

J. Trotter, African Americans in the city: the industrial era, 1900–1950. Journal of Urban History. 21 (4) (1995), 438.

D. Luca, J. Terrero-Davila, J. Stein, N. Lee, Progressive cities: Urban–rural polarisation of social values and economic development around the world. Urban Studies. 1360063X. 60 (12) (2023), 2329–2350, Available from: https://journals.sagepub.com/home/USJ, https://doi.org/10.1177/00420980221148388.

J.W. Trotter, E. Lewis, T.W. Hunter, The African American Urban Experience: Perspectives from the Colonial Period to the Present. Palgrave Macmillan, undefined Palgrave Macmillan, undefined (2004), 1–340, Available from: http://www.palgrave.com/la/book/9780312294649. https://doi.org/10.1057/9781403979162.

Alberto Vanolo, Autistic cities: Critical urbanism and the politics of neurodiversity. City. 1360–4813. 27 (1–2) (2023), 190–208, Available from: https://doi.org/10.1080/13604813.2023.2178273.

CHAPTER 11

Digital Cities

Chapter Outline

Cities today have become increasingly inseparable from the digital world of the Internet and social media. Few inventions have changed the world as much as the Internet. Used by roughly two-thirds of the world's population, the Internet has become a fundamental part of daily life, economies, interpersonal ties, education, the military, politics, and culture. Its uses include email, the most common Internet use; bill payments and electronic banking; job and housing searches; stock trading; "e-tail" shopping; searching for health information and news; online classes; digital gambling; online videogames; Voice Over Internet Protocol telephony; hotel and airline reservations; chat rooms; electronic tax payments; downloading television programs, movies, digital music, and pornography; and popular sites and services such as YouTube, Facebook, and Google. In all these ways, and more, cyberspace offers profound real and potential effects on social relations, everyday life, culture, politics, and other social activities. In this light, access to cyberspace is no longer a luxury, but a necessity. As its applications have multiplied, the Internet is having enormous impacts across the globe, including interpersonal interactions and everyday life, identity formation, and retail trade and commerce, governance, and it is affecting the structure and form of cities.

By now, digital reality and everyday life for hundreds of millions of people have become so thoroughly fused that it is difficult, if not impossible, to disentangle them. Indeed, for rapidly rising numbers of people around the world, the "real" and the virtual have become thoroughly interpenetrated. In this context, simple dichotomies such as "offline" and "online" fail to do justice to the diverse ways in which the "real" and virtual worlds for hundreds of millions are shot through with each other. Thus, rather than see cyberspace and the analog world as substitutes, we should approach them as complementary in nature.

Urban Geography. DOI: https://doi.org/10.1016/B978-0-443-30281-7.00011-6

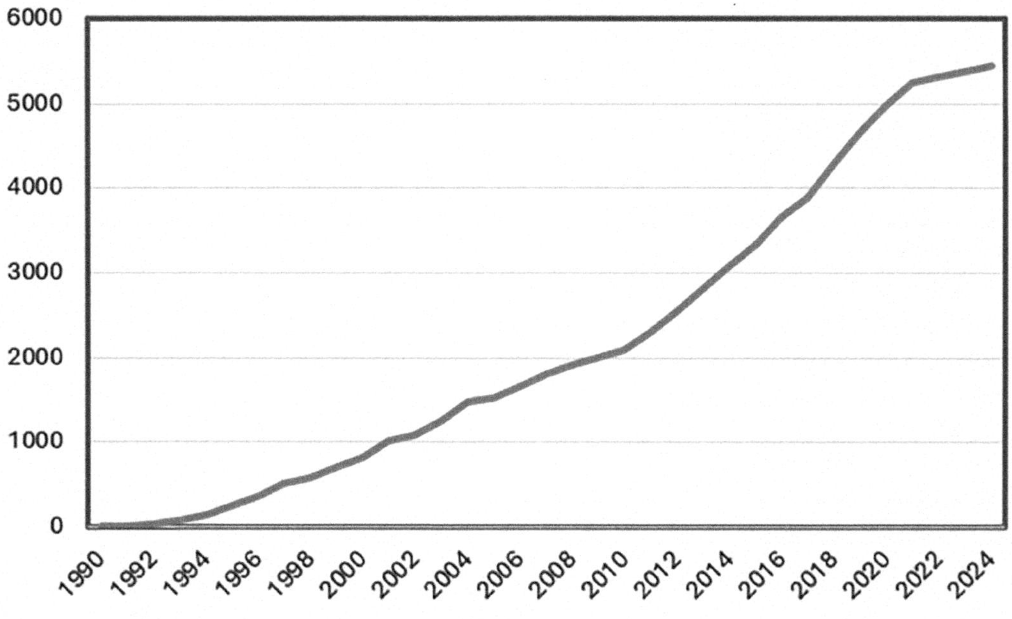

Figure 11.1
Growth in global Internet users.
(Courtesy Barney Warf.)

The growth of the Internet has been phenomenal (Fig. 11.1); indeed, it is one of the most rapidly diffusing technologies in world history. Internet usage climbed as it became clear that connected computers were more valuable than unconnected ones, as the price of equipment declined, and as cell phones made it possible to access the Internet from any location. In 2025, an estimated 5.6 billion people, or roughly 66% of the world's population, were connected to the Internet. In the 21st century, as cell phones have become widespread, the wireless Internet has become commonplace.

Although cyberspace consists of the digitized "world in the wires," and is often conceived as a domain independent of the "real" or physical world, in reality the virtual and analog worlds shape each other continuously. The Internet delivers vast amounts of services to homes, offices, and factories, including email, telephone calls, TV programs and other video images, text, and music. The Internet allowed for the rise of social media such as Facebook, WhatsApp, Instagram, Snapchat, and Twitter (now X). By now, digital reality and everyday life have become so thoroughly fused that it is difficult to distinguish them. The Internet is used for so many purposes that life without it is simply inconceivable, including email, online shopping and banking, airline and hotel reservations, videogames, chat rooms, Voice Over Internet Protocol telephony, distance education, downloadable music and television shows, blogs, YouTube, telemedicine, to simply "Googling" information. Other applications include e-government (government applications to deliver information and services) and e-commerce. Cyberspace has become a necessity for vast swaths of the world's population. Thus, simple dichotomies like "offline" and "online" do not do justice to the multiple ways in which the "real" and virtual worlds are interpenetrated. Like the printing press and the automobile, the Internet has permanently altered the ways we work, interact, consume, and live.

This chapter explores the growth and impacts of digital technologies on cities in four major steps. It begins with an overview of fiber optics cables, which are by far the most widely used telecommunications technology in the world today, and how a global network of fiber connects cities around the world. Second, it summarizes the impacts of urban digital communications, including retailing and the Internet of Things. The third section addresses urban digital divides, or discrepancies among social groups and places in terms of Internet access. Fourth, it looks at electronic governance, or e-governance, in cities. The last part addresses smart cities, which have used digital technologies to enhance their competitiveness and quality of life.

11.1 Urban Fiber Optics Infrastructures

Fiber optics lines—the seamlessly integrated network of glass wires about the size of a human hair (Fig. 11.2), bundled together in cables of several thousand—form the core of the global telecommunications infrastructure. Indeed, far more than any other technology, such as copper cables, microwaves, or satellites, fiber optics supply the vast bulk of data, voice, and video transmission services around the world. Because of their capacity to deliver high volumes of information rapidly and securely (e.g., via broadband), fiber optics cables form the backbone of the Internet as well as private corporate lines, and are widely used in the electronic media for commercial and residential purposes (e.g., cable television).

In 1977, AT&T installed the first telephone lines to use fiber optic cables, a network 25 miles in length that could carry 672 voice channels, beneath downtown Chicago; General Electric followed immediately in Boston. However, it was during the massive global changes in the world economy at the end of the 20th century that fiber came into its own as the dominant medium of telecommunications. Telephone companies and other providers of telecommunications services began rapidly replacing older copper wire cables with fiber optics, which many observers expect will become virtually the only telecommunications transmission technology in the future. Fiber optics facilitated the explosive growth of e-commerce, which includes both business-to-business transactions as well as those linking firms to their customers, including electronic data interchange (EDI) systems, digital advertising, online product catalogs, the sharing of sales and inventory data, submissions of purchase orders, contracts, invoices, payments, delivery schedules, product updates, and labor recruitment. Indeed, fiber optics arguably transformed the Internet from a communications to a commercial

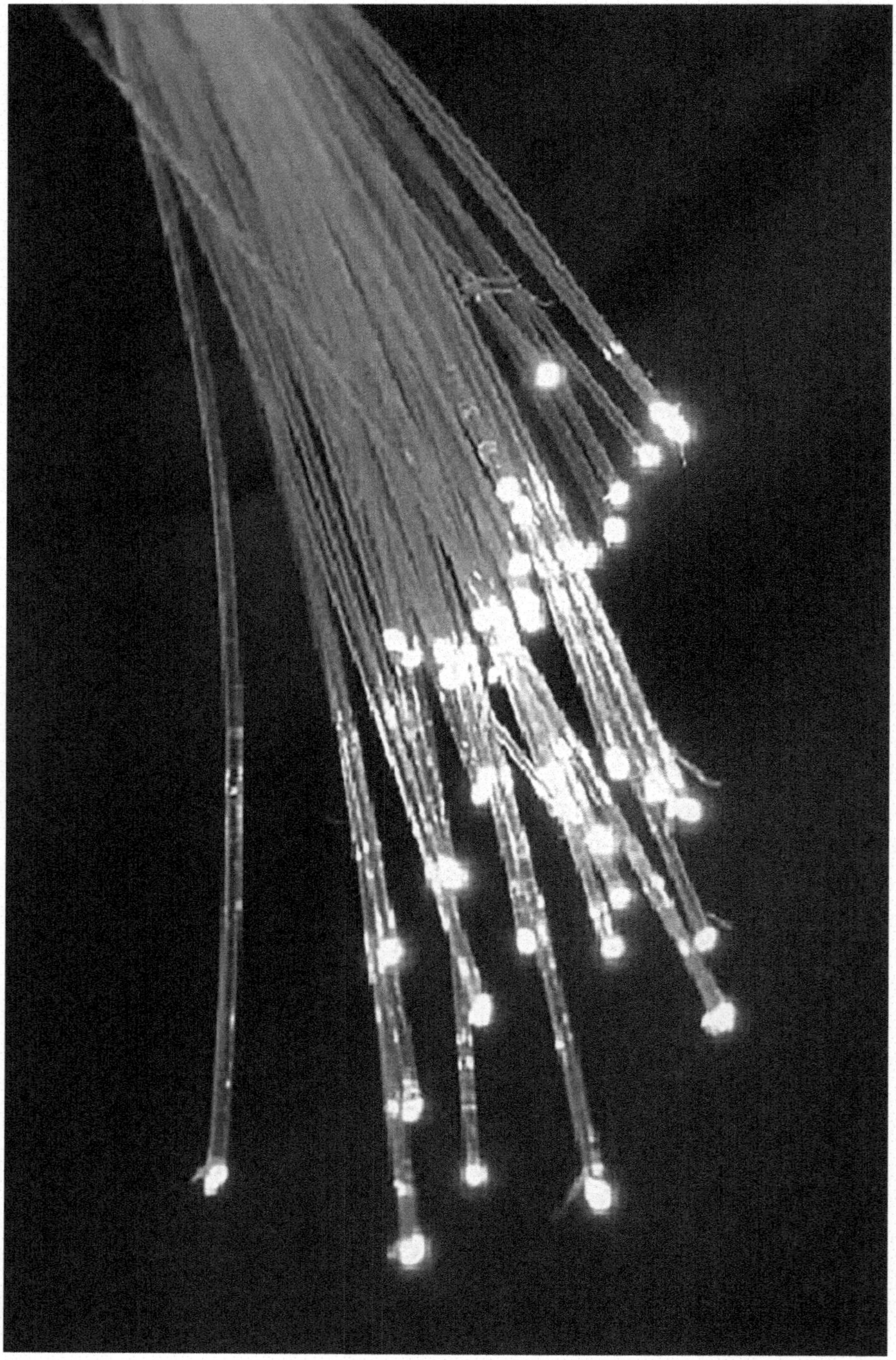

Figure 11.2
Fiber optic cables.
(https://commons.wikimedia.org/wiki/File:Fibreoptic.jpg.)

system, accelerating the pace of customer orders, procurement, production, and product delivery.

Financial and producer services firms were at the forefront of the construction of fiber networks in large part because they allowed the deployment of electronic funds transfer systems, which comprise the nervous system of the international financial economy, allowing banks to move capital around at a moment's notice, arbitrage interest rate differentials, take advantage of favorable exchange rates, and avoid political unrest. Fiber carriers are heavily favored by large corporations for data transmissions and by financial institutions for electronic funds transfer systems, in large part because of the higher degrees of security and redundancy this medium offers.

Starting in the 1980s, telecommunications firms began to build a large interurban network of fiber optics lines in the United States, whose aggregate networks exceeded 50 million kilometers. The largest fiber optics lines connect a handful of large metropolitan areas, whose comparative advantage in producer services has benefited significantly by publicly installed telecommunications systems. While the largest metropolitan regions are well served (particularly New York, Chicago, Washington, DC, Atlanta, Los Angeles, and Seattle), many other areas (such as the rural South) have few connections. High-capacity fiber lines are particularly important in regard to access to high-density material, such as graphical content on the web. For high-volume users (typically large service firms), for whom the copper cables used by telephone companies are hopelessly archaic, these lines are an absolute necessity. For large real estate developers, fiber capability has emerged as a critical issue in determining the price and attractiveness of corporate office space, indicating that relative space via connectivity is as important as accessibility via conventional transportation. Moreover, numerous cities have taken the initiative to establish their own municipal fiber networks (usually in public-private partnerships) as part of their economic development strategies to attract firms rather than wait for the private sector. In such cases, fiber lines are often packaged along with the other municipal utilities such as water, electricity, or natural gas. Thus, a grid of fiber lines surrounding the core of cities has become an indispensable part of urban comparative advantage. Rural areas, in contrast, often suffer a distinct disadvantage in this regard.

Fiber optics providers prefer large metropolitan regions where dense concentrations of corporate and residential clients allow them to realize significant economies of scale and where frequency transmission congestion often plagues satellite traffic. For example, the Atlanta metropolitan region exhibits 400,000 miles of fiber optics lines, which have been important to the revival of downtown regions and enhanced its competitive position within the national urban hierarchy. Within cities, fiber lines accelerate the creation of wealth by corporate elites, generating geographies of inequality in which the wired and the wireless, the haves and have-nots of the information age, live in close proximity; even in the most

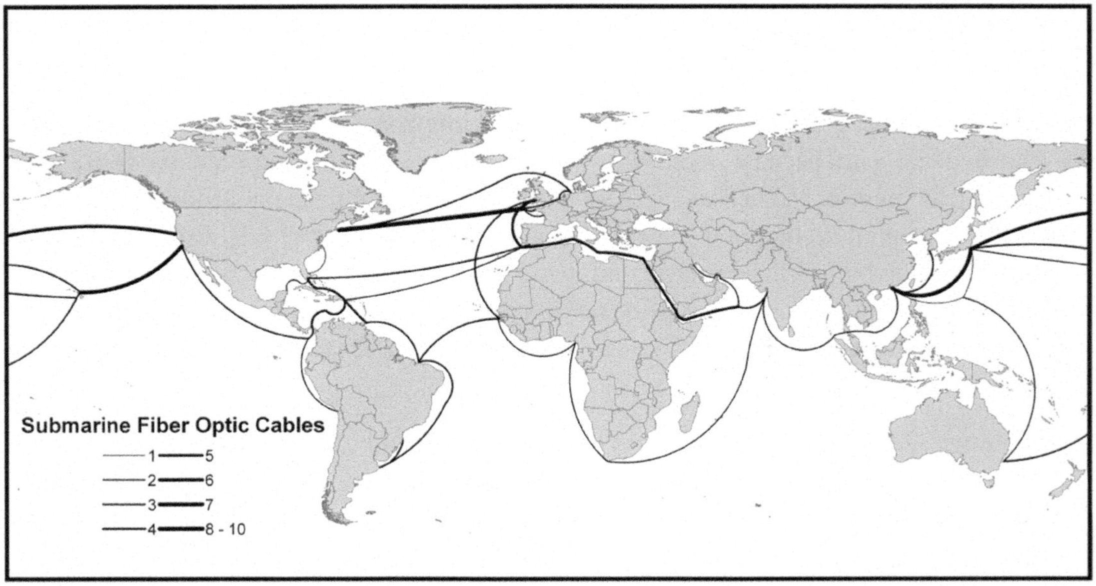

Figure 11.3
The world's major fiber optic cables.
(Courtesy Barney Warf.)

networked of cities, there exist large, disenfranchised groups who pay the costs of the digital economy but reap relatively few of the benefits. In contrast with metropolitan areas, rural areas, with relatively small populations and low market potential, hold little market appeal. This urban bias, and the social schisms it deepens, is replicated at the international scale. Despite the mythologized notion that fiber optics lines erase spatiality, it is evident that the geographic impacts of this technology are highly selective.

Today, the skein of fiber lines linking the world constitutes the core of interlinked urban systems (Fig. 11.3). Global cities such as New York, London, and Tokyo are prime beneficiaries, using fiber optics lines to spread their sphere of influence around the planet. Fiber is vital to their role and domination over the world economy.

11.2 The Impacts of Digital Communications

The urban impacts of the digital revolution are innumerable. Some have been discussed previously, such as the growth of telework and the dispersal of back offices (Chapter 4). The growth of teleworking and telecommuting has put downward pressure on office rents in large downtown areas, indicating that it is challenging the agglomeration economies there.

Cities change because of broadband networks, software applications, e-services, and datasets. These resources enable new forms of collaboration and innovation to appear. In the first place, the urban innovation system is enriched, its nodes acquire digital companions, and networking is intensified locally and extended globally. Innovation micronetworks at the city level meet the global digital networks enabled by the Internet. Digital identities facilitate innovation processes; marketplaces and crowdsourcing platforms bring additional capabilities, turning passive consumers into active producers of services; city decision-making becomes more transparent and inclusive. All these changes mark a deep transformation that affects the innovation system of cities.

The digital revolution also shaped private life in cities. The rise of streaming services such as Netflix, HBO, and Hulu creates a great diversity of entertainment options. Videogames played by people—not just kids—have become a common source of recreation. The result has been a steady withdrawal of private life inside. Children today, for example, spend much less time outside than did their peers a generation ago. The growth of home-based entertainment also entails the steady privatization of everyday life and withdrawal from communal spaces and activities, with the resulting erosion of urban social solidarity. The retreat of recreation indoors has made life more individualized and privatized, with fewer face-to-face contacts; for children, it has meant a decline in physical activity and an attendant rise in obesity.

Digital communications have profoundly reshaped urban retailing. Stores, street markets, shopping malls, and "big box" supermarkets have been instrumental in the design, growth, and social life of cities. However, since goods and services can be purchased on computers and smartphones without regard to time and location, and delivered at home a few days or even a few hours later, many stores have seen steady declines in their number of customers and revenues. Shopping malls, once ubiquitous in suburbia, have closed in large numbers.

Traditionally, the Internet is thought of as a network of computers communicating with each other. The "Internet of Things" refers to an infrastructure of ubiquitous, context-aware devices capable of communicating over the Internet. The Internet of Things includes devices such as personal computers, laptops, and mobile devices and phones that are widely used at home, as well as office devices such as printers and copy machines. Advances in wireless communications and networking technologies, together with the shrinking size and cost of microcomputers, sensors, and actuators, are driving the development of several applications such as road navigation, health monitoring, contactless purchasing, crime monitoring, and even grocery shopping. This is so because the varied interconnected devices are themselves becoming smarter, collecting a large amount of context-sensitive data automatically through their sensors without the intervention of their users. Theoretically, any device such as a refrigerator, an appliance, a car, a thermostat, or a drone, when equipped with sensing, processing, and networking capability, can act as an Internet-connected device. Some well-known examples of Internet of Things devices include self-driving vehicles, smartphones, and smart watches.

Another set of impacts of telecommunications concerns transportation informatics, including improvements such as smart metering, electronic road pricing, synchronized traffic lights, automated toll payments and turnpikes, automated road maps, information for trip planning and navigation, travel advisory systems, electronic tourist guides, remote traffic monitoring and displays, and computerized traffic management and control systems, all of which are designed to minimize congestion and optimize traffic flow (particularly at peak hours). All of these enhance the efficiency, reliability, and safety of travel. Cell phones allow more productive use of time otherwise lost to congestion.

11.3 Urban Digital Divides

Although for vast numbers of people the Internet has become a taken-for-granted dimension of everyday life, for many others—typically people with low incomes and low levels of education, older adults, and ethnic minority groups—cyberspace is an inaccessible realm. For those lacking the income to buy a computer, or who do not possess the skills to use one, or who do not have access to the Internet at school or work, cyberspace offers few benefits. Indeed, being information-deprived may cost these people opportunities such as the ability to look for jobs. People who live in a world of abundant information are those with ready Internet access; for those who do not or cannot access cyberspace, its benefits are unclear. Capitalism since the late 20th century has entered a phase in which cognitive skills and information processing are more important than ever; those deprived of Internet access pay an enormous price in foregone opportunities.

The term “access” has a variety of different definitions. It is generally taken to mean access to an Internet-connected computer at home, work, school, or public library. However, other dimensions such as physical handicaps and technical abilities are also important. Thus, equally important as access is what users *do* with the Internet, for simple access does not automatically lead one to become an Internet user. For these reasons, rather than a simple dichotomy, access should be conceptualized as a multidimensional continuum.

Significant variations exist in terms of Internet access that are unevenly distributed by wealth, gender, and ethnicity. For many economically and politically marginalized populations—again, people with low incomes and low levels of education, older adults, and ethnic minority groups– cyberspace is simply inaccessible. Access to computers linked to the Internet at home is highly correlated with income; wealthier households are far more likely to have a personal computer at home with a router and broadband than are lower-income ones. In the United States, White households use networked computers more frequently than do African American or Latino ones. Older adults likewise often find access to the Internet to be intimidating, although they are the fastest-growing demographic group of users.

Broadband has become increasingly central to Internet access. Because so much of the content of the Internet consists of large graphical files (e.g., photographs and videos), which require considerable transmission capacity, broadband access is a prerequisite to most applications. Indeed, many observers of the digital divide note that it has as much to do with access to broadband and varying uses of the Internet as it does with the simple matter of being able (or not) to log on. Thus, the broadband digital divide has become a central feature of uneven access to the Internet.

All over the world there is a sharp difference between urban and rural areas in terms of Internet access. This sharp urban-rural dichotomy reflects several forces at work. Governments often ignore rural areas in favor of politically more connected cities, especially those with global connections. In countries with deregulated and privatized telecommunications industries, Internet service providers are generally loathe to serve rural areas, whose low densities deny them economies of scale, and whose typically older, less educated, and lower-income populations do not constitute a profitable market for Internet services. Thus rural Internet connection speeds tend to be much slower than those in urban areas, and broadband is less common.

Social and spatial differentials in access to cyberspace threaten to create a largely underrepresented underclass deprived of the benefits of the Internet. Modern economies are increasingly divided between those who are comfortable and proficient with digital technology and those who neither understand nor trust it, disenfranchising the latter group from the possibility of citizenship in cyberspace. Despite the falling prices for computers, basic entry-level machines still cost more than many low-income households can afford. The recent COVID-19 pandemic illustrated how people with fewer economic resources had to resort to going to places such as McDonald's to use the Internet there; for low-income households, assignments that can be done only online are a challenge. Most public libraries offer free access to the Internet, but they are unable to serve all of those without home access. Even within the most digitized of cities there are pockets of "offline" poverty. For people without smartphones, many tasks are difficult or impossible, such as reading menus, paying for parking (Fig. 11.4), or downloading apps.

11.4 Urban e-Governance

Electronic government, or e-government, has become increasingly widespread not only in the economically developed world, but in developing countries as well. Definitions of e-government vary, but all involve the use of the Internet to deliver government information and services, change administrative procedures, and improve citizen input and participation. Web 2.0, which allows users to interact with government bureaucracies rather than just passively receive information, has expedited this process considerably. There are many types and levels of e-government, ranging from simple one-way delivery of information, in which

Figure 11.4
This sign illustrates the assumption that smartphones are ubiquitous, which is simply not the case. *(Courtesy Barney Warf.)*

citizens are passive consumers, to integration that allows user input and citizen feedback. Typically, e-government is divided into three forms: government-to-business (G2B), government-to-government (G2G), and government-to-citizens (G2C). G2B e-government includes digital calls for contract proposals; submissions of bids, bills, and payments; and Internet management of supply chains. Among other things, G2G e-government enhances interaction among different government offices and agencies. These variations are frequently alleged to increase citizen accessibility, improve efficiency, create synergies, and generate economies of scale in the delivery of government services. There are many alleged benefits of e-government. If it improves the delivery of public services, it may raise satisfaction with existing administrations; conversely, in countries plagued by chronic corruption, the increased efficiency resulting from e-government may minimize the growth of public employment and increase trust in government. Electronic payment of dues and fines limit the opportunities for graft, while digital hotlines allow citizens a voice in the circles of governance.

Many cities around the world have adopted e-government enthusiastically. In China, local governments are granted considerable autonomy in the degree to which they may implement such measures, with wide variations in the quality of websites and services. Beijing, Shanghai, Nanjing, and Shenzhen have taken the lead in online company registrations, tax collections, and community information services. For example, Shanghai has a digital social security smart card system in place, and its e-government services have become indispensable for firms operating there. Many Chinese cities have turned to social media to disseminate information; for example, many local police departments use Twitter (now X) and microblogs. Shanghai's shift towards a smart city dates back to the late 1990s. In 1998 it initiated a smart card system with many different uses, which also provides the municipal government with a centralized database containing personal information about holders (including fingerprints and medical insurance account numbers). Users can use the cards in numerous places, including hospitals and government offices, and through them can do things such as apply for home loans, drivers' licenses, marriages, divorces, and government subsidies; register their employment status in the *hukou* system; and make housing and utility payments. The system has numerous components, including the use of information technology to monitor illegal construction, street peddling, and parking, fine those who evaded paying tickets for the city's metro transit system, electronic toll collection, and dynamic parking guidance systems designed to maximize the efficiency of the city's limited supply of space. It also began implementation of programs for smart lighting, interactive electronic display boards, and a series of smartphone apps that display bus schedules as a means of discouraging driving.

Japanese cities have implemented e-government in a wide range of modes and levels of sophistication, ranging from almost nil involvement among some hamlets to intergovernmental local area networks (LANs). Seoul, South Korea, is a model of e-government. The e-Seoul safety program uses RFID tags to monitor the locations of children and people with Alzheimer's.

Because it has implemented e-government so well, Singapore has been the focus of significant attention. It has, for example, a state-of-the-art digital tax filing system and an e-citizen center, and most government services can be delivered online. The Singapore One project, which began in 1997, created the world's first nationwide broadband network, and reaches 99% of the population. A one-stop e-citizen portal, the world's first to be provided by a government, has offered a single point of access through which citizens can check traffic, download publications, register births and deaths, search for jobs, pay fees, fines, and taxes, check retirement benefits, register to vote, and obtain health care advice. The active participation of the island's business community, which anticipated its benefits, was central to the success of these initiatives. Because multiple stakeholders were included in the design and implementation of e-government initiatives, they enjoy widespread trust among different segments of the population; as with e-commerce, trust is central to the effective adoption of e-government, and trust, like all social constructions, varies greatly over time and space. Singapore has been especially innovative in using social media, particularly Facebook, to engage citizens in conversations.

In Europe, Tampere, Finland, uses e-government to allow citizen input into urban planning designs. Estonia has allowed voting over the Internet. Copenhagen uses e-government to democratize access to government records. In Greece, Trikala, the nation's first "digital city," was launched in 2006, providing near-universal Internet access and giving residents the ability to participate in telework, online library and school programs, emergency response systems, and access environmental and transportation information and demographic data that can be utilized through publicly available geographical information systems. Italian cities often use social media to inform residents of events, emergencies, and opportunities.

11.5 Smart Cities

The widespread adoption of digital technologies has given rise to the notion of smart cities, which reflects the use of information technologies for a variety of purposes. Smart cities have several components, including broadband networks (both wired and wireless), sensor networks, and various types of access devices. The smart city infrastructure includes web technologies enabling data creation, data processing, and visualization, as well as software applications in many different domains, such as entrepreneurship, education, health care, transportation, energy, public safety, and others.

Key components of smart cities include: 5 G wireless networks and next generation networks, network interoperability enabling Internet usage everywhere; mobile devices enabling ubiquitous access to data and the web; a cloud-based infrastructure, platforms and software for data storage, integration, processing, and analytics; real-world user interfaces, QR codes in signs, RFID tags, mesh sensor networks, and low energy consumption meters and control devices; applications for the web and smartphones, GPS devices, voice control, augmented reality visualization, and public data over the web, open access to data from sensors, linked data, and artificial intelligence (Fig. 11.5). Smart cities have smart buildings, which regulate

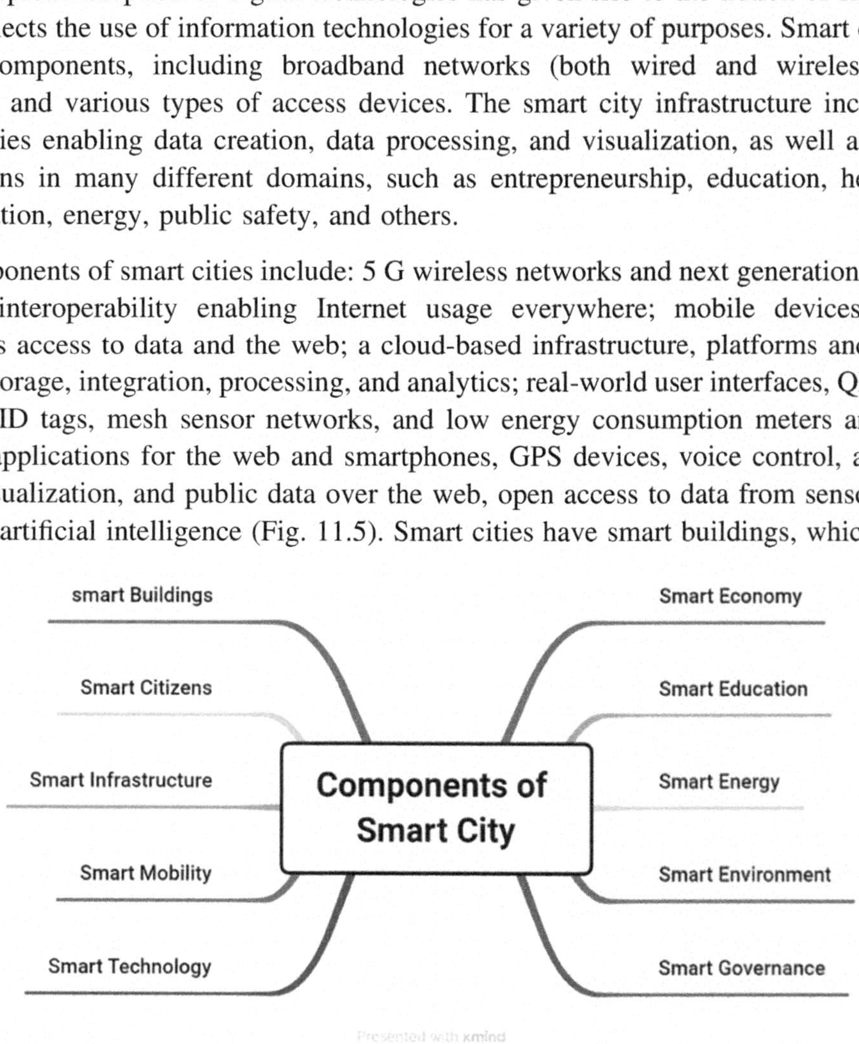

Figure 11.5
Components of a smart city.
(https://commons.wikimedia.org/wiki/File:Components_of_Smart_City.png).

temperatures depending on the presence of people and minimize energy use. Smart homes are similar, and use the Internet for a variety of purposes such as monitoring spaces, security alerts, and energy conservation.

A major component of smart cities is smart electrical grids. Over the past three decades, engineers, economists, and policymakers have been leading actions to "modernize" electricity infrastructure by adding computerized controls, sensor networks, and digital information and communications technologies. These "smart grids" retrofit electricity infrastructures with Internet-connected technologies, offering new ways of managing electricity supply and demand, new business models, social practices, and new understandings of electricity networks. Smart grid technologies include generation, transmission, distribution, and end-use applications. In the smart grid, all elements of the electrical system are transformed in some way. The smart grid entails the use of digital technologies, sensors, actuators, and other ICTs that are integrated into the existing electric grid, modernizing it to enable a variety of new capacities. The smart grid is thus a sociotechnical system that both shapes and is shaped by society, the economy, and the practices of everyday life.

Smart cities are often linked to sustainable development initiatives, including the preservation of natural habitats and ecosystems, sustainable use of land, air quality, carbon emissions, climate change adaptation, energy savings and the transition to renewable energy sources, sanitation, water management and reuse, and recycling of materials.

Examples of smart cities abound. Songdo, Korea, deploys broadband networks and sensor-based solutions for smart living. Amsterdam uses information technologies for the reduction of CO_2 emissions and has numerous smart city projects to enhance mobility, living, working, and public spaces. Barcelona deploys sensor-based solutions for parking, traffic jam avoidance, smart metering, garbage collection, environmental monitoring, and street lighting. San Francisco uses digital systems for sustainability, zero waste, recycling, CO_2 reduction, and energy conservation. Brisbane, Australia, emphasizes knowledge-based development, sustainability, and energy saving; CCTV cameras for road intelligence, smart traffic signals, and ubiquitous Wi-Fi in parks and libraries.

Seoul ranks at the top of the world's most well-connected smart cities, with one of the most developed network infrastructures on the planet. Smart meters in homes have reduced energy consumption. A citywide smart payment system allows buyers to purchase goods and services with their smartphones without entering stores—the HomePlus virtual stores allow pedestrians to make purchases using their phones and have them delivered via a mobile payments system. Seoul's "telepresence" system allows residents to conduct video chats around the world. Home devices can be programmed via mobile phones. Sensors alert riders waiting for buses as to the arrival time.

11.6 Chapter Summary

In the age of the Internet, cyberspace has become a fundamental part of how cities are structured, organized, and changed. The ubiquitous nature of information technologies at work, home, and on the street, implies that digital space and physical, analog space have become inseparable. Few urbanites could function without ready access to texts, email, or web searches.

The essential infrastructure that makes digital cities possible is fiber optics, the glass wires that transmit enormous amounts of information instantaneously. Fiber forms the backbone of the Internet and the web, and allows enormous files to be transmitted at the speed of light. The fiber infrastructure today rivals airports and highways as an essential part of urban competitiveness. A global network of fiber optics links cities across the planet, accentuating their comparative advantages under information-intensive capitalism.

The impacts of the Internet are felt not only in daily life, but in how activities in the city are organized. The Internet has given rise to new patterns of innovation. It has fundamentally altered urban retailing, as old-fashioned brick-and-mortar stores give way to "click-and-order" online services. The Internet of Things has connected enormous numbers of devices in offices and homes.

Not everyone benefits from the digital city equally; some do not benefit at all. Urban digital divides refer to inequalities within cities in terms of access to the Internet, and particularly broadband services. Because ethnic minority groups and people with low incomes have less Internet access, the digital divide reproduces in cyberspace all the inequalities found historically in cities. Everywhere across the world, however, cities have better and faster Internet connections than rural areas.

Governments, too, have embraced the Internet in the form of e-government, which allows for online searches for information; digital applications for licenses, permits, and jobs; social media alerts; and Internet payments of bills and submission of contracts. e-Government is held to streamline governments and make them more efficient and responsive, allowing for citizen feedback.

The latest chapter in this saga is smart cities, which incorporate Internet-connected devices to manage electricity, traffic, and other systems. All cities are smart cities to some extent, although some have taken the lead in deploying digital technologies. Smart cities use the Internet to deliver health care digitally and to minimize energy consumption. They have become a central tool in imagining and implementing sustainable development.

Study Questions

1. Why are fiber optics so critical to cities in the information age?
2. What causes urban digital divides?
3. What use is urban e-government?
4. What is a smart city? Give examples.

Further Reading

T. Alizadeh, Global Trends of Smart Cities: A Comparative Analysis of Geography, City Size, Governance, and Urban Planning Global Trends of Smart Cities: A Comparative Analysis of Geography, City Size, Governance, and Urban Planning. Elsevier, (2021), 1–185, Available from https://www.sciencedirect.com/book/9780128198865, https://doi.org/10.1016/B978–0–12–819886-5.01001-X.

Leo-Paul Dana, Aidin Salamzadeh, Morteza Hadizadeh, Ghazaleh Heydari, Soroush Shamsoddin, Urban entrepreneurship and sustainable businesses in smart cities: Exploring the role of digital technologies. Sustainable Technology and Entrepreneurship. 27730328. 1 (2) (2022), 100016, Available from: https://doi.org/10.1016/j.stae.2022.100016.

D. Ferreira, M. Vale, R. Miguel Carmo, L. Encalada-Abarca, C. Marcolin, The three levels of the urban digital divide: Bridging issues of coverage, usage and its outcomes in VGI platforms. Geoforum. 124 (2021), 195–206, Available from: http://www.elsevier.com/inca/publications/store/3/4/4/index.htt, https://doi.org/10.1016/j.geoforum.2021.05.002.

Stephen Graham, Global grids of glass: On global cities, telecommunications and planetary urban networks. Urban Studies. 0042–0980. 36 (5–6) (1999), 929–949, Available from: https://doi.org/10.1080/0042098993286.

Q. Hu, Y. Zheng, Smart city initiatives: A comparative study of American and Chinese cities. United States Journal of Urban Affairs. 43 (4) (2021), 504–525, Available from: http://www.tandfonline.com/loi/ujua20, https://doi.org/10.1080/07352166.2019.1694413.

Leif Johnson, Who builds Shanghai's fiber-optic network? Thinking urban infrastructure through migrant construction labor. Environment and Planning A: Economy and Space. 0308–518X. 55 (4) (2023), 795–809, Available from: https://doi.org/10.1177/0308518x221140886.

N. Komninos, The age of intelligent cities: Smart environments and innovation-for-all strategies. The Age of Intelligent Cities: Smart Environments and Innovation-for-all Strategies. Taylor and Francis Inc., Greece (2014), 1–278, Available from: https://www.routledge.com/products/9781138782198, https://doi.org/10.4324/9781315769349.

A. Kuzior, O. Pakhnenko, I. Tiutiunyk, S. Lyeonov, E-Governance in smart cities: Global trends and key enablers. Multidisciplinary Digital Publishing Institute (MDPI), Poland Smart Cities. 6 (4) (2023), 1663–1689, Available from: https://www.mdpi.com/journal/smartcities, https://doi.org/10.3390/smartcities6040078.

Abderahman Rejeb, Karim Rejeb, Steve Simske, Horst Treiblmaier, Suhaiza Zailani, The big picture on the internet of things and the smart city: a review of what we know and what we need to know. Internet of Things. 25426605. 19 (2022), 100565, Available from: https://doi.org/10.1016/j.iot.2022.100565.

T. Shelton, A. Poorthuis, M. Zook, Social media and the city: Rethinking urban socio-spatial inequality using user-generated geographic information. Landscape and Urban Planning. 01692046. 142 (2015), 198–211, Available from: http://www.elsevier.com/inca/publications/store/5/0/3/3/4/7, https://doi.org/10.1016/j.landurbplan.2015.02.020.

T. Tang, J. Hou, D.L. Fay, C. Annis, Revisit the drivers and barriers to e-governance in the mobile age: A case study on the adoption of city management mobile apps for smart urban governance. United States Journal of Urban Affairs. 43 (4) (2021), 563–585, Available from: http://www.tandfonline.com/loi/ujua20, https://doi.org/10.1080/07352166.2019.1572455.

CHAPTER 12

Conclusions

Chapter Outline

Cities are among humanity's greatest achievements, dense concentrations of people who interact in close proximity to one another. This process – in economic terms, agglomeration – has made them crucibles for human ingenuity for millennia. Cities are essentially dense nodes of interactions reflecting different forms of political economy over time. Urban areas create the vast bulk of the world's wealth, generate new ideas and products, set trends, and advance technologies. They drive the world economy.

Although they are recent compared to the very long period of hunting and gathering, cities emerged out of the Neolithic Revolution, which was marked by the invention of agriculture. The social surplus created by peasants working tirelessly in the fields sustained a new division of labor, with a ruling class, the state, and specialized occupations. Cities are where civilization was born. In myriad places across the globe, cities began to arise around 5,000 years ago, leading to a new type of society. In Mesopotamia, Egypt, the Indus River Valley, China, several places in Africa and the Western Hemisphere, and Europe, cities of varying sizes and types took shape, forever altering the course of human history. In their wake arose governments, writing, wars, taxes, and countless other things.

To understand Western cities it is necessary to appreciate their historical arc. Geographies do not appear out of thin air, they are the result of overlapping processes that often take centuries to unfold. Cities in medieval Europe, from roughly the 5th to the 16th centuries, were small, crowded, and dirty. Not surprisingly, many were devasted by the great bubonic plague of the 14th century. In feudal cities, the guilds flourished, giving rise to specialized craft workers. Nonetheless, in an age typified by the stifling rule of the Church, cities were incipient wellsprings of secular culture.

Capitalism was born in cities, stamping it with an indelibly urban character ever since. The Renaissance, the great flowering of art and science in the 15th and 16th centuries, began in Florence, in Italy. In cities, the capitalist firm found ideal locations to concentrate, compete and cooperate. Early modern cities, that is, with the onset of capitalism but before industrialization, were enriched by the colonial empires spilling out of Europe, funneling vast wealth back to the colonizing powers.

Urban Geography. DOI: https://doi.org/10.1016/B978-0-443-30281-7.00013-X

This process kicked into high gear with the onset of the Industrial Revolution in the late 18th century, when humans learned to harness inanimate energy at a large scale. With industrialization, the mechanization of labor freed countless millions from the farms, spurring rural-to-urban migration and providing a cheap source of labor for the flourishing factories. Incomes and standards of living began to rise. Industrial cities were filthy and often unsafe, lacking clean water and public sanitation. Only with the implementation of reforms in the late 19th century did they begin to shake off the layers of grime that had coated them for centuries.

With industrialization, some countries began to become predominantly urban in character. While most people lived in rural areas until the 20th century, the share living in cities has grown over time. Today, the majority of the world's population lives in urban areas.

The modern city is structured economically by the pressing demands of commodity production. Foremost among these is the need to agglomerate. The benefits accrued allow firms to be more productive when clustered together than if they were alone. This process is manifested most obviously in the Central Business District, the conglomerations of buildings that dominate downtowns. Because accessibility drives the price of urban land, the cost is highest near downtown and gradually declines with distance toward the urban periphery. This pattern shapes urban land uses as different types of land use, each capable of generating different rates of profit, vie with one another to occupy a given space.

Cities are always locked into complex webs of interaction, including supply chains, input and output linkages, trade, migration, and exchanges of information on a daily (and now, by the second) basis. When they acquire markets to export their goods and services – even in the same country – they realize revenues that generate profits, pay salaries and rents, and drive the urban economy forward.

Urban political economy is the conceptual stance that puts class and social relations front and center. One advantage of this approach is that it does not artificially divide economics from the sphere of the political. In cities, the local economy is deeply shaped by the political power of the business community. Firms often engage in alliances to shape local policies and land uses when it is to their benefit. Such growth coalitions ensure the local state is receptive to their needs, and may engage in place marketing to make their city maximally attractive to investors and tourists.

Capitalism in the late 20th century underwent a profound worldwide transformation, marked by deindustrialization in developed countries, the explosion of producer services (those sold to other firms), and rising demand for skilled, white-collar labor. The so-called "creative class" was born as a result. The massive increases in the demand for labor were accommodated by the rise of information technologies. This type of capitalism, what some call cognitive-cultural capitalism, put a premium on agglomeration, particularly that form mediated through face-to-face contact. As a result, downtown areas, long besieged by suburbanization and job loss, suddenly underwent a revival, including the emergence of massive office complexes and waves of gentrification across the planet.

Simultaneously, new telecommunications systems began to alter the world of work and urban landscapes. With the rise of the Internet, teleworking became feasible for many people, albeit only those in skilled positions that revolve around information processing. Lower-level jobs such as data entry began a steady migration to the suburbs and, in some cases, overseas as back offices became globalized. The result was a challenge to the significance of agglomeration and downward pressure on commercial property prices near the city center. These forces exemplify the continual tension between forces that draw people and firms toward the city center and those that disperse them to the periphery.

Urban economies, and cities in general, must be viewed as part of networks and systems. Even during the feudal era urban hierarchies, or webs of connected cities of varying sizes, began to take shape. Early accounts of city-systems such as Central Place Theory were analytically primitive, ahistorical, and ignored labor. More recent approaches emphasize city-regions, the large conurbations of cities and their surrounding hinterlands. The globalization of the urban hierarchy has propelled a handful of cities to the status of global cities, the command-and-control centers of the world economy, which exhibit large numbers of financial firms, stock markets, advanced corporate producer services, headquarters of transnational corporations, and serve as centers of media, fashion, and tourism. New York, London, and to a lesser extent, Tokyo, Hong Kong, Paris, and Los Angeles sit atop the world urban hierarchy.

People don't just work in cities, they live in them. The labor markets of urban capitalism generate very different incomes and residential opportunities for different groups, giving urban residential space a texture. Early attempts to come to terms with this issue, such as the Chicago School of social ecology, focused heavily on residential choice and ethnic segregation at the expense of the inevitable ties between employment and residential areas, production and social reproduction. Housing markets are complex, and reflect the intersections of many forces such as the state of the national and local economy, incomes, interest rates, population growth and decline, and public policy. Always and without exception, housing for poor people is in short supply. Similar incomes, and similar socioeconomic status and residential preferences, lead many people to live in neighborhoods with some degree of homogeneity. Neighborhoods tie residents to their communities, provide networks of assistance (e.g., crime watch programs), and are significant in the socialization of children.

Other than agglomeration, arguably the single most important force driving urban growth has been the relentless growth of the urban periphery. Suburbanization dates back to the late 19th century, but improvements in urban transportation, notably the streetcar and then the automobile, greatly enhanced the appeal and ability to live in peripheral areas. After World War II, when incomes rose rapidly and jobs were plentiful, suburbs increased exponentially in size and number. However, this process does not play out equally for all groups: in the U.S., white people have long been more able to move to the suburbs than have minorities. There are Black suburbs, but not many of them. American metropolitan areas tend to be characterized by inner city neighborhoods disproportionately populated by Black and Latino people and archipelagos of predominantly White suburbs. Suburbanization led to the political fragmentation of metropolitan areas as suburbs incorporated to protect their property tax base. Equally important, the process was devastating for many inner city areas, reducing their property tax base and the ability of cities to provide adequate public services.

The gentrification of many inner city neighborhoods has been a recurrent theme in many cities for the last half-century. Contrary to popular opinion, gentrification is not a "back to the city" movement as most gentrifiers are from the city itself. The driving force behind this process is the labor market, notably the very rapid and sustained growth of producer services creating white collar jobs. While the notion of the "creative class" reflects this process, and their lifestyles and decisions are important, they would have no income to spend were it not for the restructuring of cities around producer services that began at the same time. Gentrification is a tension-filled process, for it often entails the displacement of lower income, minority residents by upper income white ones. Indeed, it has fueled rising rents and the reconfiguration of many inner city neighborhoods, at times sparking fury and resentment among those living there.

The flip side of wealthy suburbs and gentrified communities is the urban ghetto. Ghettos are not new: the term originally applied to the Jewish quarters of European cities. Capitalism's ceaseless reproduction of uneven development, however, generated large inner city communities populated by ethnic minorities with relatively low incomes. The deindustrialization of the cores of American cities in the late 20th century was in no small part responsible for this process. With good jobs unavailable, residents face lifetimes of poverty. Many social problems faced by ghetto residents, such as high crime rates, are attributable to the lack of investment and employment there. Ghetto neighborhoods are continually threatened by poverty, unemployment, the costs of suburbanization, structural racism, and indifferent city governments.

No discussion of urban residential space would be completed without noting the depressing incidence of homelessness. In the U.S., roughly one million people are unhoused. This population disproportionately consists of people of color, and women with children are the most rapidly growing group. The blame-the-victim approach that asserts homelessness is a matter of choice is factually wrong and ignores the social context and forces that create it. Homelessness has several causes, but the most important is the lack of affordable housing: homeless people in most cases are simply down on their luck, cannot find a job paying a living wage, and caught between low incomes and expensive places to live. It is no accident that homelessness is greatest in metropolitan areas with very expensive housing, such as New York, Los Angeles, and San Francisco. Other causes include veterans experiencing psychiatric problems, the lack of mental health services, the foster care system for children, women fleeing abusive relationships, and drug addiction. There is no easy solution to homelessness, but more low-income housing and a higher minimum wage are a start.

Cities are social, economic, political, and cultural entities, but they are also environmental. They draw in vast amounts of resources, water, and energy and in turn produce copious amounts of waste. By reflecting large amounts of heat, cities are warmer than surrounding areas. Cities modify their own microclimates by creating urban heat islands. Their air is often polluted, largely due to automobile exhaust, but in developed countries has been getting slowly better. Over time, the developed world built a large and elaborate infrastructure for bringing clean water to cities, a phenomenon that greatly improved sanitation, hygiene, and urban livability. Cities must also dispose of wastewater, clean it before it returns to the environment, and deal with stormwater, which can carry oils and chemicals from streets into rivers and the seas. Urban green spaces, generally built by urban planners, offer respite from the heat and noise of the city, although not all enjoy equal access to them. The location of waste sites and of green spaces both reflect the forces of power that play out to the disadvantage of minorities: as the urban political ecology perspective emphasizes, these are questions of environmental justice. On the outskirts of cities, relentless suburbanization swallows up rural and agricultural land at a prodigious rate, a process that can be slowed by agricultural zoning.

Cities are frequently threatened by hazards of one sort or another. Those located in tectonically active areas are occasionally devastated by earthquakes, such as the one that leveled San Francisco in 1906. Heat waves periodically kill large numbers of people, particularly the elderly and vulnerable. Hurricanes, too, threaten cities, as the devastation of New Orleans by Hurricane Katrina in 2005 painfully showed, killing roughly 1,400 people. Because they have large amounts of impervious land, cities tend to have higher rates of flooding, and more severe floods, than rural areas. These "natural" hazards are anything but natural: without exception, it is the poor that suffer the most.

Climate change – the rapid heating of the Earth's atmosphere – poses an existential threat to people worldwide. Cities are also affected. Many large cities are located on the coasts and are vulnerable to rising sea levels: Rotterdam, New York, New Orleans, and Shanghai are examples. Jakarta, Indonesia, is sinking under rising sea levels, and Miami is faced with regular flooding. Climate change also exacerbates other threats to cities such as hurricanes and heat waves.

Although capitalism is often naively celebrated as an economy of "free markets" independent of government intervention, in reality the state plays a major role shaping urban environments. National urban policies are one way the government intervenes. Programs to combat poverty, protect the environment, or fund health care, housing, and transportation all profoundly shape the structure of cities and the quality of everyday life within them. So do federal tax policies. Although national urban policy in the U.S. is relatively small compared to

Europe, the federal government shapes cities by building much of the infrastructure, such as the Federal Interstate Highway system.

Urban planning, which in the U.S. is conducted at the local level, is another form of state intervention. Planning is essential in many respects, such as correcting for market failures (e.g., traffic jams, environmental destruction), preparing for emergencies such as earthquakes and floods, and monitoring the population, environment, and economy of the urban area. Planners work to preserve historical districts and agricultural lands, and minimize the impacts of urban sprawl. They help to ensure that public goods and services such as schools and hospitals are well located and deliver their outputs efficiently. Transportation planners work to expedite the flow of traffic and minimize congestion. Environmental planners help to protect wetlands and endangered species.

Planning is inescapably tied to politics. Because it is a social, not simply technical, activity, planning responds to competing interests. Real estate developers are a powerful force in municipal politics with which planners must contend. Although some pretend that planning is immune from politics, planners must work in the face of political agendas in which the priorities are not necessarily those that planners perceive to be in the public's best interest.

In the U.S., most public services are funded at the local level and provided by municipal or county governments. Two of the most important are education and transportation. Public schools are essential to an educated citizenry and social mobility, but in the U.S. there are enormous variations in the levels of funding per student and the quality of education. Typically, inner city schools are underfunded, crowded, and understaffed, while those in suburban areas are often very good. In this way the local funding of education reproduces uneven development. Public transportation usually takes the form of buses; only in large metropolitan areas with high volumes of passengers is it cost-effective to have subways.

Zoning is an important means by which planners regulate urban land use. Essentially all urban land is zoned, which is the legally permissible use of a given plot. Zoning exists to control negative externalities brought on by incompatible land uses. There are several types of zoning, including environmental, residential, commercial, and historic zoning. Zoning can also be used against people, as with exclusionary zoning, which is designed to keep certain people, that is, the poor and minorities, out of some communities using tools like minimum lot sizes.

Governments also shape cities through the provision of public housing. Because the supply of housing for the poor is chronically insufficient, governments have built public housing over the years. (They also destroy housing, as with urban renewal.) Large, decrepit public housing projects have given way to smaller ones, but usually these encounter strong local opposition in the form of NIMBY (Not In My Backyard) protests. Finally, some cities engage in rent control to protect renters from high rents, although these have the effect of discouraging the housing supply.

The bulk of the world's cities and urbanites are situated in the developing countries of Latin America, Africa, the Middle East, and Asia. There are long and rich traditions of cities in these places, but all of them were heavily influenced by colonialism as capitalism forged a global division of labor. Cities in the developing world thus emerged under very different historical conditions than did those in Europe and North America. Developing country cities are growing rapidly, driven largely by rural-to-urban migration, a reflection of the dispossession of vast numbers of peasants and an ongoing crisis in rural areas created by lack of public and private investments, poverty, and very limited opportunities. Every year hundreds of millions of people leave their villages and farms for the cities, creating huge, swollen metropolises such as Mexico City, Alexandria, Shanghai, Mumbai, Nairobi, and Rio de Janeiro.

The labor markets in developing world cities are frequently divided into the formal and informal sectors (to some extent this division is found in developed world cities as well). The formal economic sector, with relatively well-paying and stable jobs, tends to be small, including government jobs and those working for transnational corporations. In contrast, the informal sector, with a huge variety of unregulated occupations, tends to be vast, often the great majority of employment. The informal sector includes occupations such as street vendors, low-cost transportation, prostitution, poaching, recycling, and casual day laborers. Wages in the informal sector are generally low, and most people struggle to get by.

Housing in the developing world is typified by small amounts of formal housing aimed at the middle class and vast amounts of poorly constructed, self-built housing in slums and shantytowns. Typically the latter are found on the outskirts of cities and go by a variety of names: *gacekondu, kampong, favelas, colonias*, and *bidonvilles*. More than 1.5 billion people worldwide reside in such communities. While these neighborhoods vary in terms of size and income, generally their inhabitants lack ready access to clean water, electricity, public transportation, and sanitation, creating grounds for the spread of diseases such as cholera and dysentery. Government services there are often minimal. Without legal rights to the land, landowners, abetted by the government, may tear them down.

Because the developing world is vast and diverse, it is unsurprising that urbanization takes different forms in different regions. Variations in the level and nature of urbanization reflect the imprint of differing historical trajectories, roles in the global economy, national political systems, and cultural and ideological beliefs. While there are commonalities across continents, particularly poverty, urbanization in, say, Brazil is apt to be different from that in Egypt. Some developing countries are relatively wealthy, such as Malaysia; others hardly qualify as "developing" at all, such as Singapore and the United Arab Emirates, home to spectacular architecture. Geography is the study of how social processes play out in different ways in different contexts, and urbanization takes contrasting forms in different places. Cities in Latin America bear the imprint of Spanish colonialism; those in India the effects of British rule.

Cities are not simply collections of buildings and roads, but the living spaces of more than half of humanity. The social geography of urban areas is immensely important if one seeks to understand the nature of city life. The opportunities and constraints faced by urbanites vary widely depending on their class, gender, ethnicity, sexuality, age, and ability. For example, class - one's position in the urban division of labor – affords some people incomes to live comfortably in spacious homes in the suburbs or nice condominiums near the city center, while others are crowded in dilapidated apartments due to the constant shortages of low-income housing. Because class and ethnicity are always tied together, in many societies ethnic minorities have lower incomes, fewer job and housing opportunities, and often face more oppressive legal environments, such as police brutality against African Americans. Women the world over are often second-class citizens, earning less money than men and have lower access to political power. Gender structures cities by shaping the jobs in which men and women work, their incomes, commuting patterns, and disposable incomes.

Several other lines of social demarcation shape the contours of everyday life. Cities are sites of intense sexuality, especially among young adults, as reflected in the sexualized atmospheres of bars. Gay and lesbian populations, often rendered invisible for fear of discrimination, have at times formed gay communities in which they are free to express their sexuality openly. The elderly experience the city in quite different ways than do children or young adults; many elderly people have health and mobility problems, and often prefer to live near doctors and health care facilities. Disabled people, too, who include a range in the type and severity of their disabilities, often face impediments to getting from one place to another.

Cities have long had a reputation as being crime-infested places. The reality is more complex. It is true that some types of crime, such as homicide, are more common in large cities, particularly in poor areas where gangs are prevalent. Property crime rates, however, are largely invariant with respect to city size. All across the U.S., crime rates have dropped gradually, a reflection of the aging of the population and rising prosperity for large segments of the population.

Cities are not simply places of work, but also of recreation and play. For many, retail shopping is a form of recreation, although urban retailing has been dramatically changed by the Internet, much as the market for office space has been changed by telework. Recreational districts, with restaurants, theaters, and bars, cater to young adults and are busiest in the evenings. Towards late evening, however, the city becomes primarily a place for rest and sleep, the nighttime urban economy notwithstanding.

With the rise of digital capitalism in the late 20th century, the boundaries between the analog and digital worlds have become decisively blurred. The adoption of information technologies, including the Internet, relied heavily on a vast network of fiber optic cables, which have become central to urban competitiveness in an information-intensive economy. The growth of digital networks has had major impacts on urban economies, including telework, online shopping, and social interactions. However, not everyone participates equally in this new world: urban digital divides reveal how social inequalities are replicated in cyberspace. Many municipal governments have embraced digital networks to deliver services and information, allowing unprecedented citizen access. Finally, some cities have utilized information and communication technologies to become smart cities, those that deploy digital techniques to enhance the flow of traffic, conserve energy, and promote sustainable development.

How, then, should we view cities? Cities are the spatial expression of society, reflecting all the complex ways in which historical and structural forces are wrapped up in everyday life. Cities are simultaneously economic, political, cultural, and social entities filled with class conflict, ethnic diversity, and gender inequality. Cities are hugely productive and at the same time home to some of the world's most severe problems. It is critical to note these tensions and contradictions in order to appreciate urban areas in all their complexity.

Finally, cities are never static or frozen, but constantly evolve and change over time, reflecting and also contributing to broader forces such as the changing world economy, new technologies, and population changes. Cities today would be unimaginable to people living 200 years ago. The only sure thing is that cities of tomorrow will be different from those that exist today.

Further Reading

P. Adler, R. Florida, The rise of urban tech: how innovations for cities come from cities. Regional Studies. 13600591. Available from. 55 (10-11) (2021), 1787–1800, Available from: https://doi.org/10.1080/00343404.2021.1962520, http://www.tandf.co.uk/journals/carfax/00343404.html.

J. Becker, F. Chasin, M. Rosemann, D. Beverungen, J. Priefer, J.V. Brocke, M. Matzner, A. del Rio Ortega, M. Resinas, F. Santoro, M. Song, K. Park, C. Di Ciccio, City 5.0: Citizen involvement in the design of future cities. Springer Science and Business Media Deutschland GmbH, Germany Electronic Markets. 14228890. Available from. 33 (1) (2023). Available from: https://doi.org/10.1007/s12525-023-00621-y, https://www.springer.com/journal/12525.

F. Cugurullo, F. Caprotti, M. Cook, A. Karvonen, P. McGuirk, S. Marvin, Artificial Intelligence and the City: Urbanistic Perspectives on AI Artificial Intelligence and the City: Urbanistic Perspectives on AI. Taylor and Francis, Ireland (2023), 1–400, Available from: https://www.taylorfrancis.com/books/edit/10.4324/9781003365877/artificial-intelligence-city-federico-cugurullo-federico-caprotti-matthew-cook-andrew-karvonen-pauline-mcguirk-simon-marvin?context=ubx, https://doi.org/10.4324/9781003365877.

F. Cugurullo, R.A. Acheampong, M. Gueriau, I. Dusparic, The transition to autonomous cars, the redesign of cities and the future of urban sustainability. Urban Geography. 02723638. Available from. 42 (6) (2021), 833–859, Available from: https://doi.org/10.1080/02723638.2020.1746096, http://www.tandfonline.com/loi/rurb20?open=34&repitition=0#vol_34.

T.J. Dixon, M. Tewdwr-Jones, Urban futures: Planning for city foresight and city visions Urban Futures: Planning for City Foresight and City Visions. Policy Press, United Kingdom (2021), 1–290, Available from https://policy.bristoluniversitypress.co.uk/urban-futures, https://doi.org/10.46692/9781447336297.003.

Abdul Rehman Javed, Faisal Shahzad, Saif ur Rehman, Yousaf Bin Zikria, Imran Razzak, Zunera Jalil, Guandong Xu, Future smart cities: requirements, emerging technologies, applications, challenges, and future aspects. Cities. 02642751. 129 (2022), 103794, Available from: https://doi.org/10.1016/j.cities.2022.103794.

Index

Note: Page numbers followed by "*f*" and "*t*" refer to figures and tables, respectively.

Printed and bound by CPI Group (UK) Ltd, Croydon, CR0 4YY
17/09/2026
02218538-0007